SpringerBriefs in Biochemistry
and Molecular Biology

For further volumes:
http://www.springer.com/series/10196

Igor Malyshev

Immunity, Tumors and Aging: The Role of HSP70

 Springer

Igor Malyshev
Department of Pathophysiology
Moscow State University of Medicine
  and Dentistry
Moscow
Russia

ISSN 2211-9353   ISSN 2211-9361   (electronic)
ISBN 978-94-007-5942-8   ISBN 978-94-007-5943-5   (eBook)
DOI 10.1007/978-94-007-5943-5
Springer Dordrecht Heidelberg New York London

Library of Congress Control Number: 2012954380

Printed on acid-free paper

Springer is part of Springer Science+Business Media (www.springer.com)

Foreword

This book is about a very important class of biological macromolecules, namely heat shock proteins or HSP. Sometimes, these proteins are also called stress proteins. In evolutionary terms, stress proteins are perhaps the most ancient and conservative ones. They certainly were already present in the oldest ancestor of modern biological cells. Why am I so certain? Because stress proteins provide the very fundamental basis of life itself: the ability of cells to divide, turnover of intracellular proteins, cell protection from damaging factors, mechanisms of signal transduction, and gene expression. You can remove many types of proteins from cells: as a result the cell may cease to perform a function, but it will still live! But it is impossible to remove stress proteins from the cell without triggering the "death sentence" for it.

At the same time these amazing proteins were discovered very recently, only 60 years ago. They were discovered due to an error by a lab assistant working for an Italian researcher Ferruccio Ritossa. By accident the careless lab assistant increased the temperature in an incubator where Ritossa kept his *Drosophila melanogaster* flies. All laboratory staff expressed their sympathy to Ritossa for the destroyed flies and ruined experiment. However, Ritossa was not really upset. Fortunately, he was one of those excellent scientists, who look at whatever everyone looks at, but sees what nobody else sees! Due to the lab technician's error Ritossa was the first to notice that heat shock causes an increase in gene expression. These genes were called *heat shock genes*, and the proteins that they encode were called *heat shock proteins* or *stress proteins*. Thus, the accidental switching of the incubator thermoregulator "turned on" the era of heat shock proteins!

While analyzing the biological role of heat shock proteins it is certainly very important to know the details, in what concentrations they accumulate, what cell regions they work in, etc. So the English are right when they say "the devil is in the detail"! However, it is also important to understand the whole physiological picture, so the French are right as well when they say "you cannot imagine an elephant, studying it under a microscope." While looking into the issue of heat shock proteins, I tried to use both approaches: the detailed "English" and the conceptual "French" ones. The *detailed* approach means that we need to remember specific numbers, biochemical characteristics, and the sub-molecular structure of stress proteins. The *conceptual* approach means that we need to understand the mechanisms and the relationship of phenomena, to be able to see direct and reverse links, to understand, so to say, to understand the "logistics" of the intracellular management involving stress proteins. These principles underlie the writing of all chapters. It is very important to capture the connections between fundamental functions of stress proteins and their disruptions with development of specific human disease. Chapters 6–8 focus precisely on that.

For years I have been talking about stress proteins to students of the Moscow State University of Medicine and Dentistry (MSUMD), the Moscow State University (MSU), and the University of North Texas. That is why I wrote this book in the style somewhat similar to the lecture style, which is more accessible for both the students who want to discover a scientific problem for the first time, and for professionals who simply want to learn something new.

Recently, looking through my lecture notes written a decade ago, I was surprised how little we knew about stress proteins then. I am sure that ten years from now, some of the A students reading this book will think with a smile, "Poor Professor Malyshev, how little did he know!" However, I hope that, at least for today, I succeeded in describing an unbiased state of affairs in one of the most exciting areas of molecular biology and medicine, the science of stress proteins that provide the basic foundation of life.

Having finished the book, I clearly realized that I could not have written it without the help of many people. I am especially thankful to my teachers: Professor N. P. Larionov, who played a crucial role in my life by instructing me in the basics of serious scientific work; and Professor F. Z. Meerson, an outstanding world renowned scientist, who completed the "fine-tuning" of my brain on a scientific footing, and helped in choosing a subject for my doctorate thesis, suggesting to focus on stress proteins.

I am also deeply grateful to Professor V. B. Koshelev, who has been inviting me for many years to give lectures on stress proteins to students of the School of Fundamental Medicine, MSU, thus encouraging me to keep abreast of the latest works in this area. I also want to thank the professors at the University of North Texas, Drs. J. Vishwanatha, F. Downey, and E. Manukhina, who invited me to deliver a course on stress proteins to American undergraduate and graduate students; the material from these lectures is included in this book.

My University friend, Dr. D. N. Atochin, now Professor at Harvard University, was of great help in preparing the manuscript: he found a lot of great reviews and experimental reports for me. I am also grateful to Ms. Olga Rybina, a student of the School of Medicine at MSUMD for her help in finding and selecting relevant publications.

I am also grateful to all the students who attended my lectures on stress proteins. They played a big role in the decision to write this book. Sometimes, their clever and unconventional questions caught me "off guard", and often after a lecture I had to urgently dig for an answer on Internet and in electronic libraries.

I would like to express my utmost appreciation to D. N. Yushchuk, Academician of the Russian Academy of Medical Sciences and the reviewer of this manuscript, who "blessed" it to be released to the large audience.

Publishing a book is not an easy task, but all technical issues were solved thanks to the optimism and perseverance of my assistant, Ms. Anastasia Raetskaya, who is also our laboratory assistant and a student of the School of Therapy.

Finally, I have very special feelings about the invaluable support from my family including my wife, who inspired me to write this book, and my two sons, my most ardent supporters who have been impatiently waiting for the completion and publication of this book.

Contents

Chapter 1
A General Description of HSPs, The Molecular Structure of HSP70 and The HSP70 Cycle

Abstract Fifty years ago the accidental switch of a temperature knob on the incubator where Ferruccio Ritossa kept his fruit flies started a new era, the epoch of heat shock proteins (HSP). In 1986 H. Pelham was the first to suggest that HSPs bind to denatured protein aggregates, thereby restricting their aggregation and breaking them by using ATP as an energy source. Among all HSPs, the protein with a molecular weight of 70 kDa was found to be the most common, drew the most attention and is consequently the HSP we know the most about. HSP70 contains three domains: the ATPase N-domain which hydrolyzes ATP; the substrate domain which binds proteins, and the C-domain that forms the "lid" for the substrate domain. Because of its three-domain structure, HSP70 forms a unified ATPase cycle coupled with connection and disconnection of the client protein. The "team" of HSP70 cycle regulators includes **HSP40**, which delivers clients to HSP70 and stimulates ATP hydrolysis; **Hip**, which assists HSP70 in retaining the client, and **Bag-1** and **HspBP1**, which accelerate the dissociation of ADP and the release of the client protein.

Keywords HSP70 • Hip • Bag-1 • HspBP1 • The HSP70 cycle

In this first chapter, I will begin by telling you the amazing story about how heat shock proteins (HSPs) were discovered, their general characteristics, molecular structure and their functional cycle.

1.1 About the Discovery of HSP, or How *Drosophila Melanogaster* was Accidentally "Heated"

The name itself—"heat shock proteins"—results from the simple fact that HSPs were first discovered in cells exposed to elevated temperatures. As is the case with many significant scientific discoveries, they were discovered by chance. It all happened in a genetics and biophysics laboratory in the Italian city of Pavia early in the 1960s. By accident, a lab assistant increased the temperature in an incubator containing fruit

I. Malyshev, *Immunity, Tumors and Aging: The Role of HSP70*,
SpringerBriefs in Biochemistry and Molecular Biology,
DOI: 10.1007/978-94-007-5943-5_1, © The Author(s) 2013

fly *Drosophila Melanogaster*. The *Drosophila* owner, an Italian named Ferruccio Ritossa, reprimanded the lab assistant. Luckily, Ritossa was not only good at informing his lab members of bad practice, but was a good scientist too. He discovered (Fig. 1.1.) that a brief increase in temperature resulted in a characteristic pattern of puffing in the chromosomes of the salivary glands in *Drosophila Melanogaster* (Ritossa 1962). This meant that a heat shock provoked gene activation in *Drosophila*; the genes were immediately named "heat shock genes".

That in itself was one of the clearest pieces of evidence to prove that the environment influences genes and their activity. Ritossa was extremely excited; he submitted his article to one of the most prestigious scientific journals of the time. However, the editor did not accept the manuscript as "it was irrelevant to the scientific community" (Ritossa 1996). Ritossa nevertheless did not give up: he finally published in 1962 the article describing the above phenomenon in the journal *Experientia*.

However, 12 years passed before the proteins encoded by these genes were identified by the Tracy team (Tissières et al. 1974). Naturally, the proteins whose synthesis increases as a result of heat shock were called "heat shock proteins".

Finally, in 1982 Cold Spring Harbor Laboratory hosted the First International conference on heat shock. By that time the scientific community was somewhat ready to accept Ritossa's work; he was finally recognized as the discoverer of heat shock proteins.

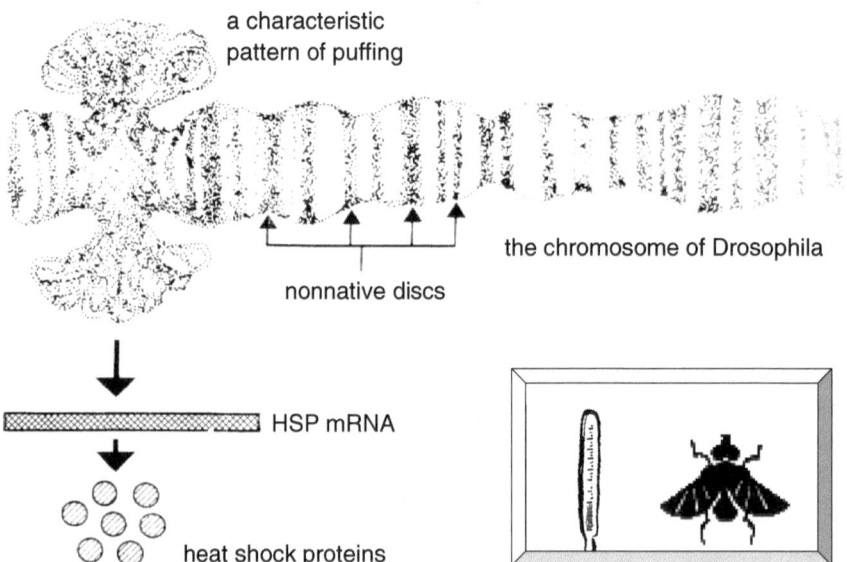

Fig. 1.1 Brief increase in temperature resulted in a characteristic pattern of puffing in the salivary glands chromosomes in *Drosophila Melanogaster*

1.2 HSPs General Characteristics: Probably the Most Conserved and Ubiquitous Proteins in all Cell Types

But what happens in cells after the heat shock? What is the "biological mission" of HSPs? Why would a cell increase the expression of HSPs? Hugh Pelham, a British scientist, was one of the first to ask, and his research gave the following results.

Immediately after heat exposure, HSP expression in a cell increases dramatically (Fig. 1.2.). Most of HSPs are located in the nucleus, or more specifically, in the nucleolus of the damaged pre-ribosomes. As time passes after the heat shock, the HSP content in the nucleus gradually decreases, while it increases in the cytoplasm. By that time, the damaged nucleolar structure and the disturbed general protein biosynthesis system have completely recovered. Is there a connection between the increase in HSP content and the recovery of general protein biosynthesis? Yes, said numerous researches who have demonstrated that inhibitors of HSP synthesis suppresses the ability of cells to restore general protein synthesis disturbed as a result of the heat shock. Therefore, no HSPs leads to no cellular recovery! This result proved that HSPs ensure the resistance of the protein synthesis apparatus to heat shocks.

However, until 1986 it was not clear how it all actually worked. An elegant hypothesis advanced by Hugh Pelham (Pelham 1986) filled in some of the gaps.

It was known that a heat shock event caused partial denaturation of protein structures in nucleolar pre-ribosomes. This process exposes hydrophobic patches that are normally buried within the interior of a protein (Fig. 1.3). The appearance of hydrophobic patches leads to protein–protein interactions. As a result, pre-ribosomes stick together to form insoluble units. That is exactly what leads to the disruption of protein synthesis in a cell. According to Hugh Pelham, HSPs first bind to the hydrophobic surfaces of the denatured proteins, thereby limiting pre-ribosome aggregation. Second, formation of a complex with denatured proteins induces the ATP activity of HSPs. Hydrolysis of ATP results in changes in the conformation of both the HSP and the bound denatured pre-ribosome protein. That in turn leads to the disruption of hydrophobic bonds between the aggregated proteins; the released proteins then get a chance to restore their initial native conformation.

Fig. 1.2 The disturbed general protein biosynthesis and HSP content in a cell after heat exposure

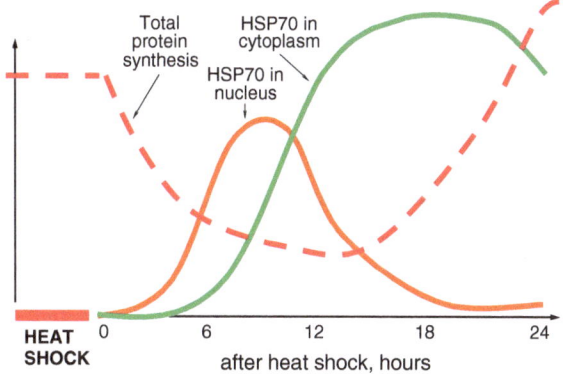

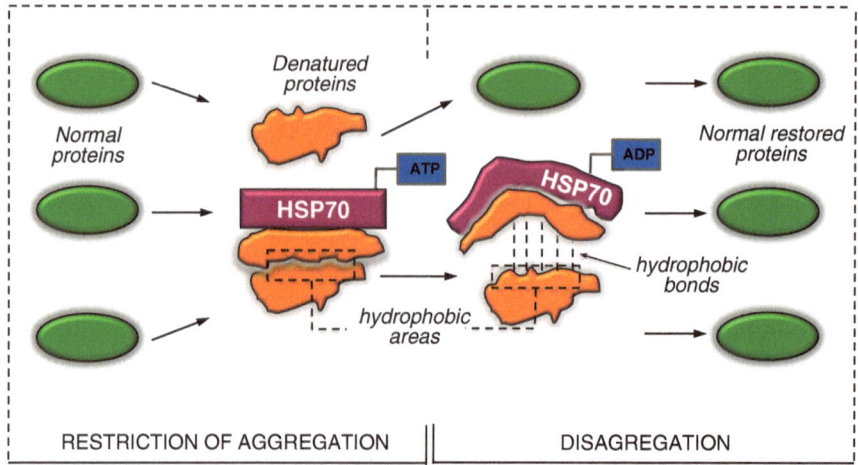

Fig. 1.3 Hugh Pelham's hypothesis on the mechanism of HSP's protective effect

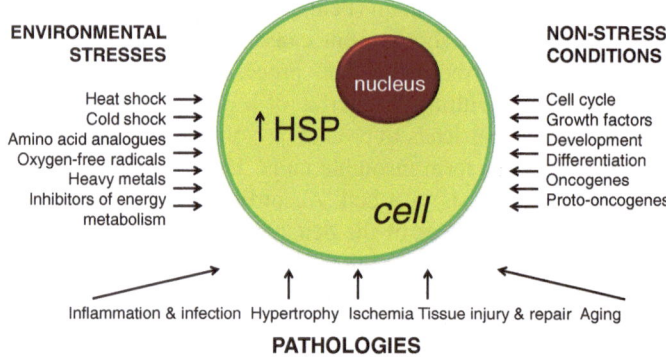

Fig. 1.4 Many different conditions and factors induce the synthesis of HSP in a cell

The sequence of such HSP-related reactions can lead to a successful disaggregation of large abnormal protein aggregates.

The basic premise of this hypothesis was that the HSP has a uniquely high affinity toward denatured proteins. Like a Spanish bull that quickly finds and attacks anything red, HSP finds and attaches to anything that has hydrophobic patches on the surface and thus inhibits protein aggregation.

After the first experiments that hinted at the important role of HSPs in protecting cells from heat shock the amount of HSP research started to snowball.

It turned out that the term "heat shock proteins" was not exactly accurate (Fig. 1.4.). As far back as 1984, Ellwood & Craig demonstrated that HSP synthesis can be equally stimulated by exposure to cold. The proteins could have been therefore easily called "cold shock proteins" or "temperature shock proteins". Then it was shown that HSP synthesis can also be induced by a number of various environmental stresses without any temperature change, as well as by various pathologies or diseases (Lis and Wu 1993; Morimoto et al. 1993; Wu et al. 1995).

With this new information, these proteins started being called "stress proteins". However, this name did not fully reflect their nature, since they were later discovered in normal cells without any stress exposure. Nowadays, most scientists worldwide still refer to them as "heat shock proteins", paying tribute to the history of their discovery.

HSPs are found in virtually all living organisms, from viruses to primates. Evolutionarily, HSPs are highly conserved proteins. For instance, homology in the amino acid composition of HSPs in bacteria and in humans reaches 60 %. By now HSPs with molecular weights of 28, 32, 40, 60, 70, 80, 90, 100 and 110 kDa have been discovered. Since tissues contain primarily the HSP with a molecular weight of 70 kDa, this HSP has attracted the majority of research attention. Therefore, we will be mostly talking about this important HSP70 family.

In 1986 Schlesinger gave a precise definition of the HSP70 family: "HSP70 family includes proteins whose synthesis is stimulated by stress, and whose genes contain a marker nucleotide block—heat shock consensus element (HSE)—CT-GAA-TTC-AG".

The HSP70 family includes at least 13 isoforms divided into two groups: constitutive and inducible. Constitutive HSP70 proteins have a high basal level and are weakly induced by stress. Conversely, inducible HSP70 are virtually absent in normal conditions, while their expression increases dramatically under stress.

In eukaryotes, HSP70 can be found in any part of a cell; in the cytoplasm, the nucleus, mitochondria, endoplasmic reticulum, as well as inside or outside any other organelle and compartment. Such a concentrated and universal spread of HSP70 indicates that they must be playing an important role in the cell.

Indeed, it was shortly found that both constitutive and inducible HSP70 isoforms were involved not only in protecting cells from heat shock, but in many other cellular processes; basically everywhere where a partial or full protein unfolding event occurs (Floer et al. 2008; Morimoto and Nollen 2002). For example, HSP70 proteins are involved in processes such as mRNA translation, protein translocation through membranes, protein delivery to sites of degradation, during the assembly and disassembly of macromolecular complexes, genes induction and apoptosis.

All these discoveries were more than sufficient to maintain the continued attention of HSP70 to molecular biologists, cell pathophysiologists and medical scientists. At the same time, as I said before, the scientific community initially failed to properly appreciate the role of these proteins. For a long time since the discovery of heat shock proteins it was poorly understood how the HSP70 protein could possibly participate in such a wide array of biological processes. Hugh Pelham's hypothesis remained an elegant one, but was still just a hypothesis. However, from the very beginning this hypothesis has established the right direction for further research by claiming that HSP70 performs its functions via reversible binding to other proteins.

1.3 The HSP70 Protein Structure: A Molecular Triptych

The "black box" of the HSP70 action mechanisms was opened by decoding the HSP70 molecular structure (Sharma and Masison 2009) It turned out that inside the cell, HSP70 exists as a dimer in which each monomer contains three functional domains: an ATPase N-domain (44 kDa), a substrate domain (18 kDa) and a C-domain (10 kDa) (Fig. 1.5). Each of the domains has its own specific function; together they form a functionally unique protein, a sort of a peculiar molecular "triptych" (meaning "a work of art containing three paintings or bas-reliefs united by a common theme or idea", from the Greek τρίπτυχος, 'three-fold').

The substrate domain functions as a binding scaffold that interacts with various substrate "client" proteins. For this purpose, the substrate domain has a special hollow. The hollow structure includes a hydrophobic internal surface and a negatively charged surrounding surface. Therefore, the substrates with an accessible hydrophobic area surrounded by positive charges have the largest affinity to the HSP70 substrate domain (Jordan and McMacken 1995).

The ATPase N-domain serves the purpose of binding and hydrolysing ATP. This function of the N-domain was determined when the molecular structure was solved. The ATP-ase N-domain of HSP70 consists of four sub-domains organized into two lobes, I and II. Lobe I is composed of sub-domains IA and IB, lobe II—of sub-domains IIA and IIB. The two lobes are connected via sub-domains IA and IIA. Sub-domains IB and IIB form a "cleft" and the bottom of this cleft forms the ATP binding site (Jiang et al. 2005; O'Brien et al. 1996). When ATP binds to the cleft, the IB and IIB sub-domains approach each other (Jiang et al. 2005).

The C-domain is composed of five α-spirals (αA-αE) and a flexible subdomain. The C-domain forms a structure similar to a "lid" on a flexible drive. Because of the flexible drive, the "lid" can open and close the substrate–binding hollow of the substrate domain (Han and Christen 2003). The "lid" covers the substrate domain and is secured or held in place by the spatial position of hydrogen and ionic bonds between the "lid" and the substrate domain. The energy of these bonds is small, so the transition between the "opened" and "closed" state of a lid can be easily implemented by simple conformational changes in the domains.

Fig. 1.5 The HSP70 protein structure: each monomer contains three functional domains: an ATPase N-domain, a substrate domain and a c-domain

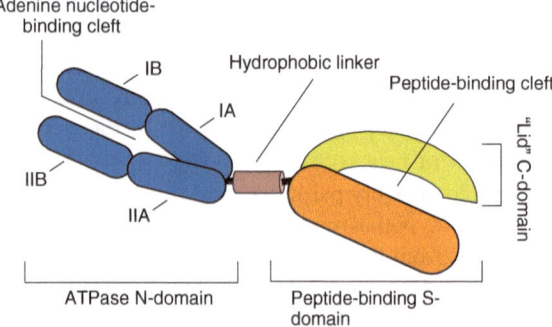

However, decoding the structure and functions of HSP70 sub-domains only partially solved the mystery of the functional mechanism of HSP70. HSP70 domains turned out to be only pieces of a whole puzzle. Sharma and Masison from the National institutes of Health (Sharma and Masison 2009) tried to piece together the complete puzzle. And they did it! The answer was that while HSP70 performs its functions, its domains are actively interacting with each other. Moreover, these inter-domain interactions form a single ATPase cycle associated with the attachment and detachment of the substrate protein. Figure 1.6 presents the HSP70 cycle.

1.4   The HSP70 Cycle: The Mechanism of the ATP-Dependent Interaction of HSP70 with Other Proteins. Or What is Common Between the HSP70 Cycle and the Flight of a Bumblebee

Binding of the substrate protein to the substrate domain of HSP70 gives rise to changes in the substrate domain conformation (Fig. 1.6). This change represents a signal that is transmitted to the N-domain and induces ATP hydrolysis. ATP hydrolysis in turn causes a change in the N-domain conformation. The change in the N-domain generates two types of conformational call-back signals. The first signal is transmitted to the substrate domain: this leads to an increase in binding affinity of the substrate to HSP70. The second signal is transmitted to the C-domain. This signal leads to the C-domain "lid" covering the hydrophobic hollow of the substrate domain, thus "attaching" the substrate to HSP70 even more.

Fig. 1.6   The HSP70 cycle: the mechanism of the ATP-dependent interaction of HSP70 with other proteins

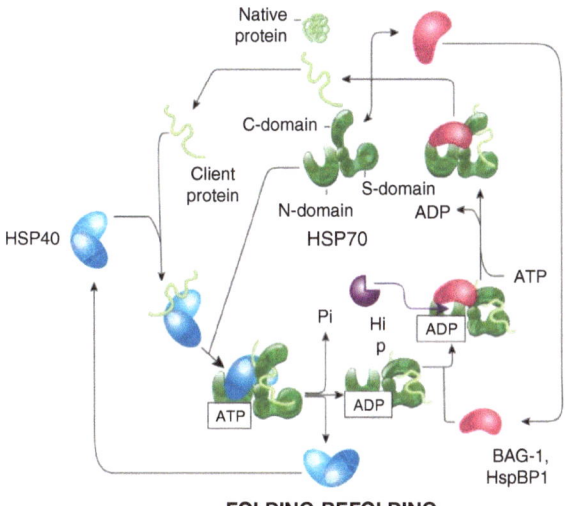

FOLDING-REFOLDING

Therefore HSP70 in the ADP-conformation retains the substrate protein because the sub-strate domain has high affinity to the substrate and the C-domain "lid" covers the bound substrate.

The next stage of the ATPase cycle includes removal of ADP and the binding of ATP into the cleft between the IB and IIB sub-domains of the N-domain. The binding of ATP to the N-domain generates two conformational signals, opposite to those generated by ATP hydrolysis. The first signal, induced by ATP binding, is generated between the sub-domains IA and IIA in the N-domain and is transmitted to the substrate-binding hollow through the substrate domain spiral. This reduces the substrate affinity of the substrate domain. The second conformational signal is also formed in the N-domain; however, this signal is transmitted to the C-domain whereby conformation changes lead to the C-domain opening its "lid". As a result, nothing holds the substrate protein anymore and it detaches from HSP70.

Therefore, HSP70 in the ATP-conformation does not retain the substrate protein because the substrate domain has low affinity to the substrate and the "lid" of the C-domain is opened. In this conformation, HSP70 is ready to attach a new protein and to start a new ATPase cycle.

This model could have explained the mechanism of HSP70 activity, if not for one observation. The ability of HSP70 to release ADP and to hydrolyse ATP is very low. For example, the rate of ADP dissociation is only 3–4 molecules per minute while the rate of ATP hydrolysis is up to 1 molecule per minute (Ha et al. 1999; Brehmer et al. 2001). With such a low rate of the ATPase cycle HSP70 would barely be able to perform any biologically significant task.

This situation reminds us of an old story about German aviation scientists. Once they wanted to create a flying engine based on the principles of bumblebee flying. Having calculated all aerodynamic characteristics of the bumblebee wings they realized that the bumblebee with such parameters would not be able to fly! However, the bumblebee did not know about those scientific calculations and kept flying. Likewise, HSP70 continued functioning perfectly.

1.5   Bag-1, HspBP1, Hip and HSP40: "helper" Proteins for HSP70, but Each One is Canny

The ability of HSP70 to release ADP and hydrolyse ATP was evaluated in vitro, whereas HSP70 works in a live cell and perhaps the rate of release and hydrolysis differ. Therefore it was suggested that cells may have special molecular instruments to accelerate the HSP70 ATPase cycle. Indeed, soon such instruments were discovered by Dr. Bracher from the Max Planck Institute of Biochemistry in Germany (Shomura et al. 2005) and Brodsky JL from Pittsburgh University (Brodsky et al. 2002).

These two scientists found that the low rate of nucleotide exchange at the HSP70 N-domain could be substantially increased by special protein factors. They were called Nucleotide Exchange Factors (Brodsky et al. Brodsky et al. 2002a, b; Dragovic et al. 2006). In higher eukaryotes such factors are Bag-1 and HspBP1 (Shomura et al. 2005; Brodsky et al. 2002a, b).

It turned out that Bag-1 interacts with the cleft between sub-domains IB and IIB of the HSP70 N-domain (Harrison et al. 1997; Sondermann et al. 2001) and that it induces the IIB sub-domain to rotate by 14°. This small sub-domain IIB holds ADP and such a rotation liberates ADP (Fig. 1.6). Because of this simple mechanism Bag-1 increases the ADP dissociation rate more than 600 fold (Gassler et al. 2001).

Therefore, we should say a few words about this important regulator. In humans the Bag protein family consists of six homologs: Bag-1, Bag-2, Bag-3, Bag-4, Bag-5 and Bag-6 (Takayama and Reed 2001). All the homologs contain a Bag domain, which is required for the interaction with the released HSP70 (Takayama et al. 1999; Miki and Eddy 2002). However, only Bag-1 is able to exchange nucleotides for HSP70. In addition to the Bag domain, Bag-1 contains other domains that can interact with other proteins involved in different cell processes. This is very important for understanding the variety of HSP70 functions. For example, Bag-1 contains an ubiquitin-like domain. Due to the presence of these two domains Bag-1 can simultaneously bind to both HSP70 and a proteasome, thereby coupling the ATPase HSP70 cycle to proteolysis of the released substrate in proteasomes (Lüders et al. 2000; Gassler et al. 2001).

For now it is important to remember that Bag-1 is a protein factor that speeds up the dissociation of ADP from the HSP70 N-domain, thus controlling the substrate release from HSP70.

At the same time, it is well known that the principle of biological regulation is often based on the balanced action of activators and inhibitors. Consequently, when an ADP dissociation activator was discovered, scientists immediately asked the question: are there also any inhibitors of this process in a cell? This turned out to be a valid question. In 1995, Hohfeld, Minami and Hart, researchers from the Howard Hughes Medical Institute and the Sloan-Kettering Cancer Center (USA) (Hohfeld et al. 1995) discovered the Hip protein (molecular weight of 43 kDa). Hip was found to interact with the ATPase domain of human HSP70 (Höhfeld et al. 1995). Currently Hip is considered an antagonist of Bag-1 (Fig. 1.6.). Hip competes with Bag-1 for binding on the ATPase domain of HSP70 to prevent the Bag-1-stimulated release of the ADP nucleotide (Kanelakis et al. 2000; Lambert and Prange 2003) and, consequently, the release of the immature substrate (Höhfeld and Jentsch 1997).

Another rate-limiting step of the cycle is that the rate of ATP hydrolysis can be substantially increased by the HSP70 interaction with co-factors such as HSP40 (Lu and Cyr 1998; Wegele et al. 2003). As I have already discussed, the binding of the substrate stimulates ATP hydrolysis only slightly; this is not sufficient for closing the C-domain lid and triggering the whole HSP70 cycle. HSP40 provides assistance (Fig. 1.6.) by interacting with HSP70 via its J-domain (Karzai and McMacken 1996; Laufen et al. 1999) and thus increases the ATP hydrolysis rate more than a 1,000 fold (Liberek et al. 1991; Laufen et al. 1999). As mentioned above, ATP hydrolysis leads to strong capture uptake of the client protein by the HSP70 substrate domain. It was clearly demonstrated that the ability of HSP40 to stimulate this mechanism was important for virtually all HSP70 activities. (Kelley 1999).

Therefore we need to say a few words about this very important HSP70 partner. Unlike HSP70, HSP40 has many more isoforms. Humans only have 13 HSP70 isoforms while HSP40 has 41 isoforms. Like HSP70, Hsp40 functions as a dimer. HSP40 can also bind hydrophobic peptides, and can independently prevent protein aggregation (Lian et al. 2007; Moriyama et al. 2000). However, the main physiological destiny of HSP40 is to assist HSP70 in performing its functions. HSP40 can interact with virtually all molecular and sub-molecular participants of the HSP70 cycle including N- and C-domains and the substrate protein.

Furthermore, it appears that the ability of HSP40 to increase the ATP hydrolysis rate at the HSP70 N-domain was only one of many "services" that HSP40 provides to HSP70. In fact, in addition to the J-domain, HSP40 has several variable domains. Some of these domains can interact with different substrate client proteins and deliver them to HSP70 (Demand et al. 1998; Landry 2003). Other domains of HSP40 can interact with different protein structures of organelles and cellular structures. Therefore, when HSP40 binds to HSP70, first of all it allows HSP70 to capture a much broader spectrum of protein substrates than it would have done without the assistance of HSP40 (Misselwitz et al. 1998). Second, HSP40 aids the localization of HSP70 in various places inside the cell, such as in the cytosol close to ribosomes, that allows HSP70 to participate in protein translation from cytoplasmic ribosomes (Horton et al. 2001); in mitochondria or the endoplasmic reticulum which allows HSP70 to participate in the transfer of the polypeptide chain across the membrane of these organelles (Brodsky et al. 1998) or where biogenesis of peroxisomes takes place (Hettema et al. 1998). Third, as we have mentioned, it increases the rate of ATP hydrolysis in the HSP70 N-domain.

Therefore, HSP40 substantially expands the field of activity of HSP70 (Genevaux et al. 2007) and ensures incorporation of HSP70 into various cell processes in numerous regions of a cell.

Now let us look again carefully at the whole ATPase HSP70 cycle. Although at a first glance it does not seem to be important, note the following: HSP40 controls the rate of ATP hydrolysis, and the delivery and binding of the client protein, whereas other factors like Bag-1 and Hip control the rate of ADP dissociation and the release of the client. You can truly appreciate it and admire how elegantly and efficiently the cell utilizes its resources if you understand that the nucleotide exchange factors and HSP40 in a cell can be spatially separated. This means that the substrate protein can associate with HSP70 in one place, where HSP40 is available, and dissociate in an absolutely different location, where there is the nucleotide-exchange factor Bag-1 but no Hip! And this is no less than the intercellular system of vector protein delivery!

At this place my students usually shout "So we can actually use this system to deliver medicines to a specific cell region?" There is no doubt that many pharmacologists would love to obtain a driving license for such a vehicle!

1.5.1 *To Sum it up, or What we Have Learnt (Summary)*

To conclude, we can summarize what those who have never heard about heat shock proteins would have learnt by reading this chapter the following points:

1. Fifty years ago an accidental switch of a temperature knob on the incubator where Ferruccio Ritossa kept his *fruit flies* started a new era, the epoch of heat shock proteins (HSPs).
2. HSP synthesis can be activated by various physical, chemical and biological stressors.
3. In 1986, Pelham was the first to suggest that HSPs bind to denatured protein aggregates, restrict protein aggregation and break protein aggregates using ATP as an energy source.
4. Starting with the work of Pelham a great deal of data have been accumulated about the fundamental role of HSPs in the survival of cells.
5. Among all heat shock proteins, the species with a molecular weight of 70 kDa was found to be the most common, thereby drawing the largest amount of attention from researchers and is consequently the one we know the most about.
6. HSP70 contains three domains: the ATPase N-domain, which hydrolyses ATP, the substrate domain, which binds proteins, and the C-domain that forms the "lid" for the substrate domain.
7. Because of its three-domain structure, HSP70 forms a unified ATPase cycle coupled with the association and the disassociation of the client protein.
8. The "team" of HSP70 cycle regulators includes: **HSP40,** which delivers clients to HSP70 and stimulates ATP hydrolysis; **Hip,** which assists HSP70 in retaining the client; and **Bag-1** and **HspBP1,** which accelerate ADP dissociation and the release of the client protein.

In the next chapter we will examine the HSP70 cycle in more detail and how it is involved in the most important intracellular processes such as protein folding, protein transportation into organelles, and directing old or incorrectly folded proteins for degradation. We will also examine the role of the HSP70 cycle in signal mechanisms.

References

Brehmer D, Rüdiger S, Gässler CS et al (2001) Tuning of chaperone activity of Hsp70 proteins by modulation of nucleotide exchange. Nat Struct Biol 8(5):427–432

Brodsky JL, Kabani M, McLellan C et al (2002a) HspBP1, a homologue of the yeast Fes1 and Sls1 proteins, is an Hsc70 nucleotide exchange factor. FEBS Lett 531(2):339–342

Brodsky JL, Kabani M, Beckerich JM (2002b) Nucleotide exchange factor for the yeast Hsp70 molecular chaperone Ssa1p. Mol Cell Biol 22(13):4677–4689

Brodsky JL, Lawrence JG, Caplan AJ (1998) Mutations in the cytosolic DnaJ homologue, YDJ1, delay and compromise the efficient translation of heterologous proteins in yeast. Biochemistry 37(51):18045–18055

Demand J, Lüders J, Höhfeld J (1998) The carboxy-terminal domain of Hsc70 provides binding sites for a distinct set of chaperone cofactors. Mol Cell Biol 18(4):2023–2028

Dragovic Z, Broadley SA, Shomura Y et al (2006) Molecular chaperones of the Hsp110 family act as nucleotide exchange factors of Hsp70 s. EMBO J 25(11):2519–2528

Floer M, Bryant GO, Ptashne M (2008) HSP90/70 chaperones are required for rapid nucleo-some removal upon induction of the GAL genes of yeast. Proc Natl Acad Sci USA 105(8):2975–2980

Gassler CS, Wiederkehr T, Brehmer D et al (2001) Bag-1 M accelerates nucleotide release for human Hsc70 and Hsp70 and can act concentration-dependent as positive and negative cofactor. J Biol Chem 276(35):32538–32544

Genevaux P, Georgopoulos C, Kelley WL (2007) The Hsp70 chaperone machines of Escherichia coli: a paradigm for the repartition of chaperone functions. Mol Microbiol 66(4):840–857

Jiang J, Prasad K, Lafer EM, Sousa R (2005) Structural basis of interdomain communication in the Hsc70 chaperone. Mol Cell 20(4):513–524

Jordan R, McMacken R (1995) Modulation of the ATPase activity of the molecular chap-erone DnaK by peptides and the DnaJ and GrpE heat shock proteins. J Biol Chem 270(9):4563–4569

Han W, Christen P (2003) Interdomain communication in the molecular chaperone DnaK. Biochem J 369(Pt 3):627–634

Ha JH, Hellman U, Johnson ER et al (1999) Structure and mechanism of Hsp70 proteins. In: Bukau B. (ed.) Molecular chaperones and folding catalysts. Regulation, cellular function and mechanism. Harwood Academic Publishers, Amsterdam, pp 573–607

Harrison CJ, Hayer-Hartl M, Di Liberto M et al (1997) Crystal structure of the nucleotide exchange factor GrpE bound to the ATPase domain of the molecular chaperone DnaK. Science 276(5311):431–435

Hettema EH, Ruigrok CC, Koerkamp MG et al (1998) The cytosolic DnaJ-like protein djp1p is involved specifically in peroxisomal protein import. J Cell Biol 142(2):421—434

Höhfeld J, Jentsch S (1997) GrpE-like regulation of the hsc70 chaperone by the anti-apoptotic protein BAG-1. EMBO J 16(20):6209–6216

Höhfeld J, Minami Y, Hartl FU (1995) Hip, a novel cochaperone involved in the eukaryotic Hsc70/Hsp40 reaction cycle. Cell 83(4):589–598

Horton LE, James P, Craig EA, Hensold JO (2001) The yeast hsp70 homologue Ssa is required for translation and interacts with Sis1 and Pab1 on translating ribosomes. J Biol Chem 276(17):14426–14433

Kanelakis KC, Murphy PJ, Galigniana MD et al (2000) Hsp70 interacting protein Hip does not affect glucocorticoid receptor folding by the hsp90-based chaperone machinery except to oppose the effect of BAG-1. Biochem 39(46):14314–14321

Karzai AW, McMacken R (1996) A bipartite signaling mechanism involved in DnaJ-mediated activation of the Escherichia coli DnaK protein. J Biol Chem 271(19):11236–11246

Kelley WL (1999) Molecular chaperones: how J domains turn on Hsp70 s. Curr Biol 9(8):R305–R308

Lambert C, Prange R (2003) Chaperone action in the posttranslational topological reorientation of the hepatitis B virus large envelope protein: implications for translocational regulation. Proc Natl Acad Sci U S A 100(9):5199–51204

Landry SJ (2003) Structure and energetics of an allele-specific genetic interaction between dnaJ and dnaK: correlation of nuclear magnetic resonance chemical shift perturbations in the J-domain of Hsp40/DnaJ with binding affinity for the ATPase domain of Hsp70/DnaK. Biochemistry 42(17):4926–4936

Laufen T, Mayer MP, Beisel C et al (1999) Mechanism of regulation of hsp70 chaperones by DnaJ cochaperones. Proc Natl Acad Sci U S A 96(10):5452–5457

Lian HY, Zhang H, Zhang ZR et al (2007) Hsp40 interacts directly with the native state of the yeast prion protein Ure2 and inhibits formation of amyloid-like fibrils. J Biol Chem 282(16):11931–11940

Liberek K, Marszalek J, Ang D et al (1991) Escherichia coli DnaJ and GrpE heat shock proteins jointly stimulate ATPase activity of DnaK. Proc Natl Acad Sci U S A 88(7):2874–2878

Lis J, Wu C (1993) Protein traffic on the heat shock promoter: parking, stalling, and trucking along. Cell 74(1):1–4

Lüders J, Demand J, Höhfeld J (2000) The ubiquitin-related BAG-1 provides a link between the molecular chaperones Hsc70/Hsp70 and the proteasome. J Biol Chem 7:4613–4617

Lu Z, Cyr DM (1998) Protein folding activity of Hsp70 is modified differentially by the hsp40 co-chaperones Sis1 and Ydj1. J Biol Chem 273(43):27824–27830

Miki K, Eddy EM (2002) Tumor necrosis factor receptor 1 is an ATPase regulated by silencer of death domain. Mol Cell Biol 22(8):2536–2543

Misselwitz B, Staeck O, Rapoport TA (1998) J proteins catalytically activate Hsp70 molecules to trap a wide range of peptide sequences. Mol Cell 2(5):593–603

Morimoto RI, Kroeger PE, Sarge KD (1993) Mouse heat shock transcription factors 1 and 2 prefer a trimeric binding site but interact differently with the HSP70 heat shock element. Mol Cell Biol 13(6):3370–3383

Morimoto RI, Nollen EA (2002) Chaperoning signaling pathways: molecular chaperones as stress-sensing 'heat shock' proteins. J Cell Sci 115(Pt 14):2809–2816

Moriyama H, Edskes HK, Wickner RB (2000) [URE3] prion propagation in Saccharomyces cerevisiae: requirement for chaperone Hsp104 and curing by overexpressed chaperone Ydj1p. Mol Cell Biol 20(23):8916–8922

O'Brien MC, Flaherty KM, McKay DB (1996) Lysine 71 of the chaperone protein Hsc70 is essential for ATP hydrolysis. J Biol Chem 271(27):15874–15878

Pelham HR (1986) Speculations on the functions of the major heat shock and glucose-regulated proteins. Cell 46(7):959–961

Ritossa F (1962) A new puffing pattern induced by temperature shock and DNP in Drosophila. Experientia 18:571–573

Ritossa F (1996) Discovery of the heat shock response. Cell Stress Chaperones 1(2):97–98

Sharma D, Masison DC (2009) Hsp70 structure, function, regulation and influence on yeast prions. Protein Pept Lett 16(6):571–581

Shomura Y, Dragovic Z, Chang HC et al (2005) Regulation of Hsp70 function by HspBP1: structural analysis reveals an alternate mechanism for Hsp70 nucleotide exchange. Mol Cell 17(3):367–379

Sondermann H, Scheufler C, Schneider C et al (2001) Structure of a Bag/Hsc70 complex: convergent functional evolution of Hsp70 nucleotide exchange factors. Science 291(5508):1553–1557

Takayama S, Reed JC (2001) Molecular chaperone targeting and regulation by BAG family proteins. Nat Cell Biol 3(10):E237–E241

Takayama S, Xie Z, Reed JC (1999) An evolutionarily conserved family of Hsp70/Hsc70 molecular chaperone regulators. J Biol Chem 274(2):781–786

Tissières A et al (1974) Protein synthesis in salivary glands of Drosophila melanogaster: relation to chromosome puffs. J Mol Biol 84(3):389–398

Wegele H, Haslbeck M, Reinstein J, Buchner J (2003) Sti1 is a novel activator of the Ssa proteins. J Biol Chem 278(28):25970–25976

Wu X et al (1995) The effect of hypothermia on induction of heat shock protein (HSP)-72 in ischemic brain. Metab Brain Dis 10(4):283–291

Chapter 2
The Functions of HSP70 in Normal Cells

Abstract In normal cells, the HSP70 ATPase cycle performs several fundamental functions: (1) together with co-chaperones, HSP70 forms a protein folding mechanism and provides protein transportation into organelles; (2) assisted by HSP40, HSP70 recognizes irreversibly damaged proteins and, assisted by CHIP, Bag-1 and HSJ1 ubiquitinates these proteins, thereby targeting them for degradation via proteasomes; and (3) together with the co-chaperones HSP90, HSP40, Hip, Hop and Bag-1, HSP70 recognizes normal proteins containing the marker sequence KFPRQ and sends these proteins for degradation in lysosomes. Thus, the HSP70 ATPase cycle forms a protein quality control system or the FOlding Refolding Degradation machinery (FORD) and, depending on the state of the protein, sends the protein either for re-folding or for degradation. Because of the FORD machinery, a cell maintains protein homeostasis. The HSP70 ATPase cycle also controls the activity of key signalling proteins by maintaining these proteins in an inactive or active state by regulating their levels and by intracellular transport.

Keywords HSP70 • Protein folding • HSP90 • CHIP • Proteasomal degradation • Lysosomal degradation

In the first chapter, I introduced an important class of proteins that exist in cells—heat shock proteins. Synthesis of these proteins, primarily HSP70, dramatically increases when a cell falls into adverse conditions. HSP70 can quickly receive the signal that damaged proteins and abnormal protein formations have appeared in a cell and repair the damage. In this sense, HSP70 acts as an intra-cellular "911" service!

As soon as the first hints about the protective role of HSP70 appeared in scientific journals, scientists started asking: what is the structure of these proteins? How do they act? Significant efforts went into answering these questions. As a result, we now know that HSP70 contains three domains: the ATPase N-domain that hydrolyzes ATP, the substrate domain that binds proteins, and the C-domain that provides a "lid" for the substrate domain. We know now that, because of its three-domain structure, HSP70 forms a single ATPase cycle that involves the association and disassociation of the client protein. We also know that the "team" of

HSP70 cycle regulators include: **HSP40**, which delivers the clients to HSP70 and stimulates ATP hydrolysis; **Hip**, which assists HSP70 in retaining the client, and **Bag-1** and **HspBP1**, which increase the rate of ADP dissociation and the release of the client protein.

The idea of the ATPase cycle of HSP70 provided an explanation as to how HSP70 can protect a cell from heat damage. According to the Pelham hypothesis, HSP70 binds to denatured protein aggregates, restricts their aggregation and breaks protein aggregates using the energy generated from the hydrolysis of ATP. A fairly logical system of ideas on HSP70 was built: these ideas regarded HSP70 as a protective protein which is rapidly synthesized "on demand" in response to cell damage.

However, HSP70s are also found in normal cells. Consequently, what would be the role of HSP70 in this situation? How does the ATPase cycle assist HSP70 in performing its functions in a normal cell? This is the subject of the second chapter.

It is impossible to define who began thinking about the role of HSP70 in a normal cell. However, the one who started addressing this problem in the right direction certainly knew two things. This "Mister X" knew, first of all, that HSP70 could bind to the hydrophobic part of proteins and delay protein folding. Second, he knew the problem in explaining the protein folding phenomenon. In fact, in vitro most proteins fold spontaneously, confirming the Anfinsen principle implying that a polypeptide chain sequence contains all the necessary information to define the exact three-dimensional protein fold (Anfinsen 1973). However, the situation in a cell is much more complicated. The cytoplasm is full of various molecules and organelles, and the problems arise already during the translation: growing polypeptide chains may begin aggregating with each other. (Dobson 2003). Such events happen because the hydrophobic patches in these chains stay exposed until the entire polypeptide chain comes off the ribosome. It is only after the synthesis is completed that the polypeptide chain can fold, burying all hydrophobic regions of the polypeptide in the interior of the protein.

Having compared these two points, it was easy to propose that HSP70 may be the best candidate for playing a role in assisting protein folding, because it can bind and shield hydrophobic zones of the polypeptides being synthesized, thus preventing unwanted protein aggregation. Once the entire chain is synthesized, HSP70 can release the chain, allowing it to fold properly and to reach its native three-dimensional state.

However, even after a proper folding event, proteins tend to aggregate, since small physical and chemical changes within the cell or just normal functioning may expose hydrophobic regions thereby leading to protein aggregation. Here, HSP70 may play a substantial role in preventing the aggregation of normal proteins.

Therefore, by analyzing the role of HSP70 in a normal cell, researchers have noted first of all that, in terms of functionality, HSP70 was well suited to solving problems that arise with the appearance of hydrophobic regions, both in the process of translation and protein folding, and during the normal functioning of mature proteins.

2.1 The Role of HSP70 in Protein Folding and the Prevention of Protein Aggregation

So how does HSP70 guarantee correct protein folding in normal cells? In eukaryotes, HSP70, in cooperation with its co-factors, interact with at least 30 % of all synthesized proteins to ensure that the correct native conformation is achieved. It all starts at the point where HSP40 assists each emerging hydrophobic area of a protein synthesized on the ribosome to bind to the substrate domain of HSP70 (Fig. 2.1.). The events we discussed in the previous chapter then occur (Hartl and Hayer-Hartl 2009). HSP40 initiates ATP hydrolysis within the N-domain. This increases the protein affinity of the substrate domain; at the same time the flexible C-domain "covers" the bound protein from the top. Now we understand why it is important that HSP70 binds to hydrophobic areas of the polypeptide chains! Shielding of the hydrophobic areas by HSP70 prevents unwanted aggregation of protein chains. Removing HSP70 gives rise to the situation where the cell will have a significant population of proteins that will not form functionally active tertiary structures. During polypeptide chain synthesis, factors such as Hip keep hydrophobic areas bound to HSP70. Further, when the whole polypeptide chain emerges from the ribosome and is ready

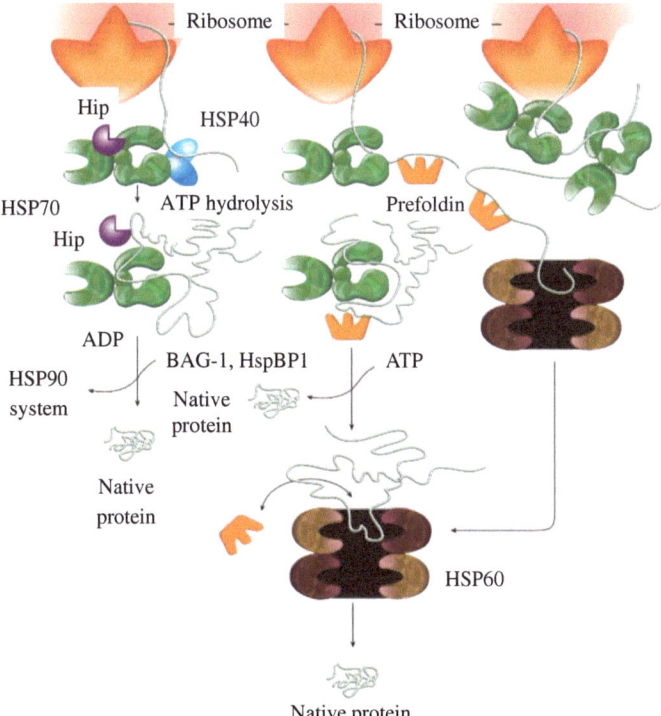

Fig. 2.1 How HSP70s guarantee correct protein folding in the cytoplasme of normal cells

to fold into a native tertiary structure, the nucleotide exchange factors exchange the N-domain ADP for ATP. As you remember from the first chapter, this leads to a reduction in protein affinity toward the substrate domain, opening of the "lid" and eventually to the release of the entire polypeptide chain and its folding.

In addition to HSP70, HSP40 and nucleotide exchange factors, in some cases, the process of protein folding may involve HSP60—barrel-shaped particles, inside which the protein folding process takes place. Protein barrels further isolate synthesized polypeptides or denatured proteins to ensure favourable folding or refolding of proteins. HSP60 are involved in the folding of ~10 % of all proteins.

Some proteins do not fold in the cytoplasm, but in the endoplasmic reticulum (Fig. 2.2.). For this process, the protein synthesized on a ribosome should be transported into the reticulum through a translocon, and the first to meet the "guest" protein is HSP70, which is called BiP in the endoplasmic reticulum. BiP binds to the first protein hydrophobic area to appear in the ER and thus prevents the polypeptide chain from returning back into the cytoplasm. In addition, with the help of repeated ATPase cycles, BiP binds every time to the next hydrophobic area of the protein which has just appeared in the reticulum lumen, and pulls the polypeptide chain into the lumen of the reticulum like a ratchet. Interestingly, the function of HSP40 in attracting BiP and regulating its ATPase activity is performed by translocon proteins located on the endoplasmic reticulum membrane.

Proteins destined for mitochondria also bind to HSP70 rather than folding in the cytoplasm. The function of cytoplasmic HSP70 is to maintain a linear state of the protein chain, so that the protein can penetrate through the narrow translocons of the mitochondrial membranes (Fig. 2.3.). As soon as the beginning of the

Fig. 2.2 How HSP70/BiP guarantee correct protein folding in the endoplasmic reticulum of normal cells

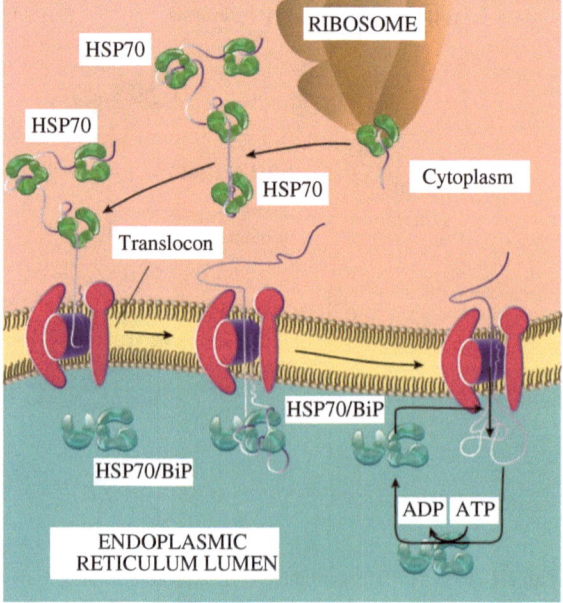

Fig. 2.3 The role of HSP70 in protein folding in the mitochondria of normal cells

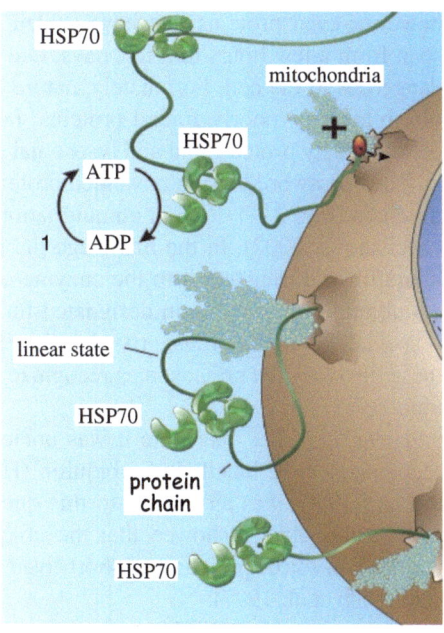

polypeptide chain appears in the matrix, other mitochondrial HSP70 would, by the energy of ATP, draw the protein chain into the mitochondrial matrix. In the same location, within the matrix there are regulators required for the cycle: the nucleotide exchange factor and translocon proteins accelerate the ATPase activity of mitochondrial HSP70. Retraction of the protein into the mitochondria occurs in the same way as explained above for the events occurring in the endoplasmic reticulum.

Therefore HSP70 carefully escorts polypeptide chains from the moment of its birth on a ribosome until the final formation of the protein spatial structure. *Since HSP70 largely does what a nanny does for children, HSP70 proteins were named "chaperones"* (Bukau et al. 2006). *Chaperone, according the Oxford dictionary, means an older married or widowed woman who accompanies a young woman to her first ball (or a person who accompanies and looks after another person or group of people or an older woman responsible for the decorous behaviour of a young unmarried girl at social occasions).*Proteins that help chaperones to perform their functions, such as HSP40, were named co-chaperones.

2.2 The Role of Hsp70 in Protein Degradation: Welcome to an Execution!

Unfortunately, errors sometimes occur during protein folding and re-folding. This can lead to the emergence of non-functional proteins or to proteins that are harmful to the cell. Folding of mutant proteins leads to the same consequences;

however, even proteins that have folded properly have a limited functional life-span from a few hours to a few days. Old dysfunctional proteins can even become dangerous to the cell. Fortunately, nature provided several ways of disposing such old, mutant or poorly folded proteins; heat shock proteins participate in two of them, namely proteasomal and lysosomal degradation.

The "password" for an unwanted protein entering into a proteasome is a ubiqui-tin chain (Fig. 2.4.). Protein ubiquitination consists of three stages (Glickman and Ciechanover 2002). In the first stage the ubiquitin-activating enzyme E1 binds to ubiquitin and transfers it to the enzyme-carrier E2. In the second stage E2 deliv-ers ubiquitin to the protein designated for degradation, and then in the third stage a special E3 ligase transports ubiquitin from E2 to the "wrong" protein. After these three events, proteasomes recognize and degrade the proteins "labelled" with ubiquitin.

However, for a long time it was unclear how damaged and old proteins were recognized to be labelled by ubiquitin (Hershko and Ciechanover 1998). Only in the mid-1990s was an answer to this question found. Goldberg and Ciechanover laboratories clearly showed that the ubiquitin–proteasome system needs HSP70 and HSP40 for degradation of short-lived, mutant or old proteins (Lee et al. 1996; Bercovich et al. 1997).

It turned out that HSP70, assisted by its co-chaperone HSP40, recognizes dam-aged and irreversibly denatured proteins and sends them to proteosomes (Cyr et al. 2002) with the help of co-chaperones such as CHIP, Bag-1 and HSJ1 (Lüders et al. 2006; Petrucelli et al. 2004).

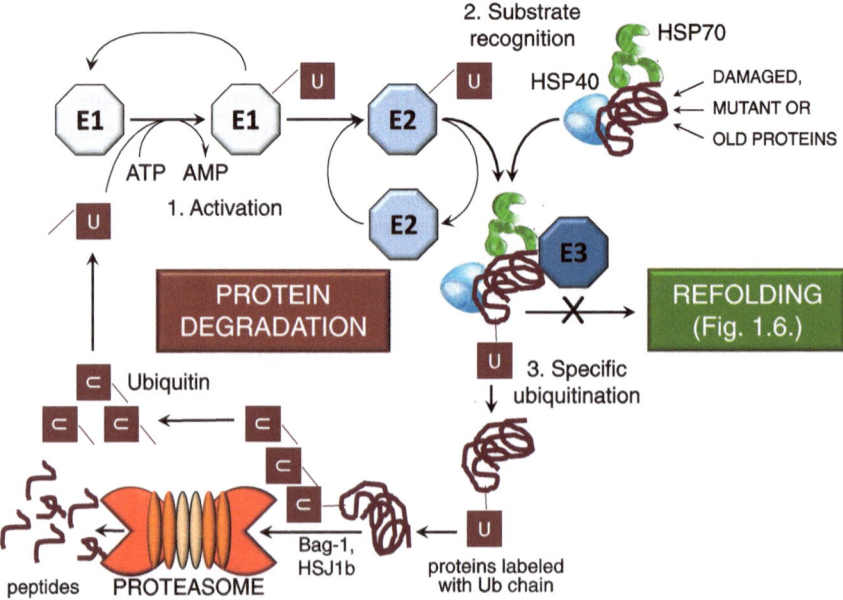

Fig. 2.4 The role of Hsp70 in protein degradation into proteasomes

The exact mechanism remains unclear; however, when HSP70 binds an irreversibly denatured or abnormal protein, the HSP70 C-domain exposes the sites for binding the co-chaperone **CHIP** (C terminus of HSP70 interacting protein) (Ballinger et al. 1999; Connell et al. 2001). This is a key moment in the switching of the effector function of the HSP70 ATPase cycle from the folding function to the proteolytic function (Fig. 2.4.). This is where the fate of the client protein is decided: whether there is a refolding process or whether the denatured protein is sent for proteasomal degradation. So why does CHIP play such an important role?

Ballinger et al. (Ballinger et al. 1999) discovered and described properties of CHIP that allowed the understanding of the mechanism of chaperone-dependent ubiquitination. CHIP was found to be an E3 ligase (Ballinger et al. 1999; Connell et al. 2001) and it has binding sites not only for HSP70 or HSP90, but also for the E2 enzyme-carrier (Jiang et al. 2001). CHIP binds to HSP70 and thereby interferes with the ATPase cycle and suppresses the HSP70 capacity for protein refolding (Ballinger et al. 1999); however, at the same time it builds the «HSP70-HSP40-protein-client-CHIP-E2» complex. When this complex is built, CHIP ubiquitinates the client protein bound to HSP70, thereby labelling the protein for proteasomal degradation (Connell et al. 2001; Glover and Lindquist 1998; Meacham et al. 2001).

Ubiquitin labelling is sufficient for the protein to enter the proteasome and be degraded. There are, however, two more co-chaperones, Bag-1 and HSJ1b, which until the very end control the execution of the ubiquitinated protein.

You know already that **Bag-1** contains a BAG domain, which is necessary for the interaction with HSP70. You also know that Bag-1 acts as a nucleotide exchange factor for HSP70, causing the release of the client protein. Finally, by now you also know that this Bag-1 activity is necessary for normal folding of the client protein. Since the ATPase cycle can only function in one of the two alternative states, it seems obvious that the folding activity of the cycle should inhibit its proteasomal activity. Indeed it was demonstrated that Bag-1 does inhibit the proteasomal degradation of the Tau protein[1] because of an increase in the refolding activity of the HSP70 cycle (Elliott et al. 2007).

However, nothing is that simple. The "double game" of Bag-1 is that in addition to the BAG domain it also contains an ubiquitin-like domain, through which it can interact with proteasomes (Lüders et al. 2000) and transport the ubiquitinated protein into proteasomes (Alberti et al. 2003). This is, for example, how the glucocorticoid receptor gets degenerated (Demand et al. 2001).

While the ubiquitinated client protein is Bag-1-escorted to its proteasome "scaffold", there is always a danger that ubiquitin hydrolases would cleave the ubiquitin chain off the client protein, and the protein would no longer be recognized for proteasomal degradation. The system of the "sentenced execution" for the disposal of unwanted proteins takes care of this danger too. To make sure it

---

[1] Tau proteins stabilize microtubules in neurons of the central nervous system. When tau proteins are defective, they can result in Alzheimer's disease.

does not happen, another co-chaperone, **HSJ1b**, which contains ubiquitin-binding sites, binds to ubiquitinated proteins and protects them from ubiquitin hydrolases, thus ensuring that the ubiquitinated proteins are delivered to proteasomes. (Westhoff et al. 2005).

Membrane proteins and proteins of the endoplasmic reticulum lumen are directed for proteasomal degradation through a mechanism called endoplasmic reticulum-associated degradation (ERAD). In this mechanism, the reticular HSP70, Bip, recognizes some poorly folded proteins in the reticulum (Molinari et al. 2002) and binds directly to these mis-folded proteins (Nishikawa et al. 2005; Zhang et al. 2001). BiP, assisted by other proteins, then transports the denatured protein to the inner side of the reticulum membrane and transfers it through the membrane to cytoplasmatic HSP70. HSP70 then delivers these proteins to proteasomes for degradation.

An alternative way of degrading intracellular proteins—the one assisted by lysosomes (Fig. 2.5.)—comes to the foreground when a cell exists in a state of prolonged fasting.

Using the culture of human fibroblasts Chiang (Chiang et al. 1989) showed that HSP70 recognizes the protein destined for degradation, binds to it and transports it to the lysosome.

Majeski and Dice (Majeski and Dice 2004) helped to establish details of this process. When a cell is starving interesting things happen: normal proteins start degrading. This apparently "strange" cell behaviour is caused by the necessity to supply the protein biosynthesis process with amino acids to produce proteins essential to the survival of the cell. It turns out that up to 30 % of normal proteins sacrificed by a cell and sent to lysosomes contain a marker amino acid sequence known as **KFPRQ**. This degradation "black mark" is recognized by the

Fig. 2.5 The role of Hsp70 in protein degradation into lysosomes

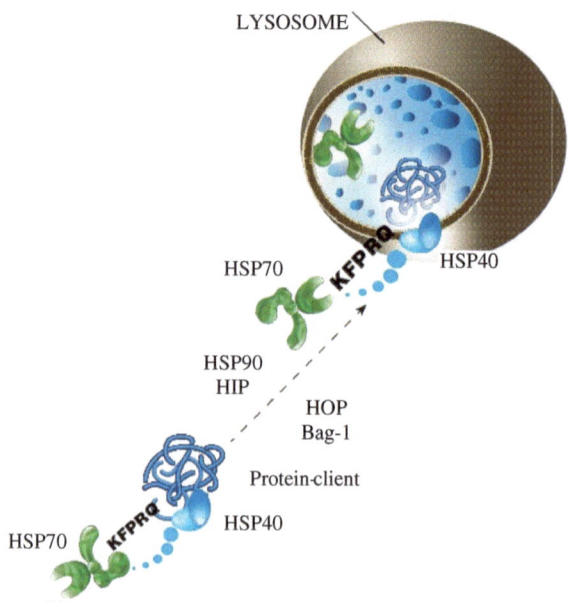

HSP70/HSP90 chaperone complex that also contains co-chaperones Hsp40, Hip, Hop and Bag-1. This complex then binds to the lysosome membrane and imports the client protein to the lysosome to be degraded (Salvador et al. 2000).

This is quite incredible! When the cell was only being formed and experienced no starvation, it was already defined at the genetic level (by the **KFPRQ** motif) what proteins would be sacrificed in the situation of possible starvation.

Another intriguing fact was that HSP70 contains two KFERQ sequence motifs. Thanks to these motifs HSP70 penetrates into the lumen of lysosomes, where it is probably involved in protein transportation. However, lysosomes do not contain ATP; therefore the lysosomal HSP70 cannot retract proteins into the lysosome lumen using the ATP-dependent ratchet, similar to what mitochondrial HSP70 and Bip do in the lumen of the endoplasmic reticulum. The type of mechanism the lysosomal HSP70 use remains unresolved.

So we have examined *the role of HSP70 in proteasomal and lysosomal protein degradation*. A careful comparison of these two mechanisms of proteolysis shows something important: *in both cases HSP70 together with its co-chaperones performs similar functions, namely, it recognizes the "unwanted" protein and delivers it to the disposal site: to either proteasomes or lysosomes*. It appears also that the choice of the disposal site is determined genetically by specific amino acid motifs, such as KFPRQ.

Another important conclusion we can make is that the HSP70 ATPase cycle—together with chaperones and co-chaperones which ensure folding and re-folding, and the proteasomes or lysosomes that ensure degradation of "irreparable" proteins—form the protein "quality control" mechanism (Fig. 2.6.). To make it easier to remember the effector function of the molecular chaperone machine, one can use the abbreviation **FORD-machinery** to define the whole protein quality control system. It is made of the first letters of the words *Folding–Refolding–Degradation*, namely the **FORD machinery**.

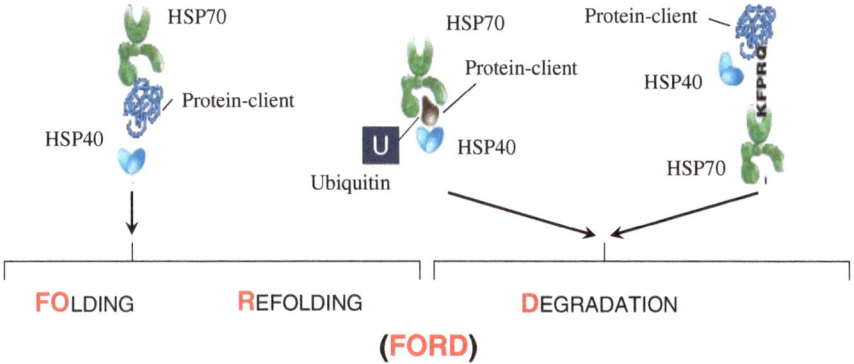

Fig. 2.6 The HSP70 ATPase cycle which ensures folding and re-folding, and the proteasomes or lysosomes that ensure degradation of "irreparable" proteins form the protein "quality control" mechanism (FORD-machinery)

2.3 Folding–Refolding–Degradation: The Molecular Protein Quality Control Machinery. The One that Decides the Fate of a Protein

Since HSP70 interacts with a number of intercellular proteins, its participation in folding or degrading should be carefully controlled. An important question arises then. Who switches the FORD-machinery on (or off)? At least two mechanisms play a potential role in forming a switch.

The first switch mechanism is formed as a result of competition between folding-stimulating co-chaperone and the factors that stimulate protein degradation for HSP70 binding sites. Normally the concentration of the folding-stimulating co-chaperones is 5–10 times higher than the concentration of the degradation stimulating factors. (Connell et al. 2001; Demand et al. 1998; Höhfeld and Jentsch 1997) Therefore under normal conditions, the main role of HSP70 will be protein folding. However, under various influences on a cell, the ratio of different co-factors may change, and that can switch the cycle from the folding-refolding pathway to degradation.

We know the examples of both co-factors. CHIP, and sometimes Bag-1 and HSJ1 can stimulate degradation. In humans, the nucleotide-exchange factor HSPBP1 is the activator of HSP70-mediated protein folding and, consequently, the degradation inhibitor. (Esser et al. 2004). HSPBP1 can bind HSP70 and CHIP to hamper protein ubiquitination (Alberti et al. 2004). As a result, HSPBP1 can switch the HSP70 cycle from degradation to folding. It was shown, for instance, that HSPBP1 inhibits the CHIP-mediated Cystic fibrosis transmembrane conductance regulator (CFTR)[2] protein degradation (Alberti et al. 2004; Arndt et al. 2005).

Another mechanism of switching the functional activity of the HSP70 ATPase cycle arises from chaperone HSP90. Whitesell and coworkers were the first to conduct research that made scientists think about the role of HSP90 in regulating the functions of HSP70 (Whitesell et al. 1994) and Sepp-Lorenzino et al. (Sepp-Lorenzino et al. 1995). These two groups showed that inhibition of HSP90 by the antibiotics herbimycin A and geldanamycin significantly increased the ubiquitination and proteasomal degradation of certain proteins. The simplest explanation is that HSP90 inhibits the CHIP-dependent ubiquitination of the client protein and can switch the HSP70 cycle from degradation to folding.

In this way, HSP70 manifests itself as a sort of molecular transformer: when certain factors get attached, the HSP70 cycle ensures folding of the newly produced or denatured and damaged protein; when other factors are attached, the cycle does the opposite; the protein is targeted for degradation.

Taken together, this allows **HSP70** with its co-chaperones and the chaperone HSP90 to build an intracellular system of protein quality control. In this system, HSP70 acts as a "judge" together with its "jury" of co-chaperones to decide the

---

[2] CFTR is an ion channel that transports chloride ions across epithelial cell membranes. Mutations of the CFTR gene affect functioning of the chloride ion channels, leading to cystic fibrosis.

fate of a damaged or denatured protein. Here, a protein will be either sent for "re-education", i.e., refolding, or the protein will be sent for degradation to the proteasomal or lysosomal "scaffold".

2.4 The Role of HSP70 in Signal Transduction

The concept of HSP70 and its co-chaperones' role in coupling mechanisms of protein folding and degradation also made researchers revise or supplement their understanding of the mechanisms for intracellular signal transduction. Each year additional data show that normal intracellular activity of signalling proteins is controlled by HSP70. These proteins include, for instance, steroid hormone receptors, Raf kinase, eIF2α-kinase, CyclinB1/Cdk1 and heat shock transcription factors–1 (HSF-1), c-Myc and pRb.

Through its interaction with such signalling molecules, HSP70 is involved in cell cycle regulation, differentiation and transduction of hormonal signals. It is not surprising therefore that HSP70 and its co-chaperones play an important role in the process of normal development as well as in pathological processes, such as carcinogenesis, aging and neurodegenerative dysfunctions (Jäättelä 1999; Kregel 2002).

A few general principles of the effect of HSP70 on signal transduction mechanisms were discovered.

First, HSP70 complexes with a signalling client protein also involve various co-chaperones, such as HSP40, CHIP, chaperone HSP90 and some other proteins (Pratt 1997).

Second, generally in such complexes the intracellular transduction proteins remain in an inactive state from which they can be quickly activated (Pratt and Toft 2003). For instance, HSP70 detachment from the complex, when denatured or damaged proteins appear in a cell, can lead to activation of signalling proteins, because HSP70 has a high affinity towards denatured proteins. Therefore adverse environmental conditions can activate intracellular transduction pathways. This is how activation of the HSF-1 transcription factor occurs.

In a second example, HSP70 together with HSP90 keeps the intracellular steroid receptor in an inactive form that is still ready to bind a ligand. When the steroid hormone binds to the receptor, the chaperone complex disassociates, and the active hormone-receptor complex is directed to the nucleus and activates gene transcription.

Third, HSP70 and its co-factors can interact with components of signalling pathways to protect them also from degradation and to ensure their continued active state. HSP70 employs this "set of tactics" with respect to the retinoblastoma protein (pRB). pRB is a major regulatory molecule involved in the initial stages of the cell cycle. HSP70 protects pRB from degradation and allows active pRB to restrain the cell from entry into the cell cycle.

Fourth, through rapid switching of its folding and degradation activities the ATPase cycle can effectively regulate the quantitative and qualitative composition of a certain part of the pool of signalling molecules. This can also ensure effective regulation of signalling pathways. For instance, hypoxia-induced factor I (HIF-I) is a transcription factor that mediates the cell adaptation to hypoxic conditions. Under normal conditions, HSP70 and CHIP interact with HIF-1 and thus ubiquitinate this protein and direct it for proteasomal degradation (Luo et al. 2010). Under hypoxia, denatured proteins appear in a cell. This leads to HSP70 detachment from HIF-1 and attachment to the damaged proteins, to which it has a higher affinity. As a result, HSP70 stops sending HIF-1 for degradation and HIF-1 can penetrate into the nucleus and activate genes responsible for adaptation to hypoxia.

Finally, HSP70 can be involved in the intracellular trafficking of signalling proteins. This was clearly shown in the example of G protein-coupled receptors (GPCRs). In order to transduce an extracellular signal into the cell GPCRs have to be properly folded, transported to the cell surface and subsequently degraded once they have performed their function. The HSP70 chaperone FORD-machinery plays the major role in organizing all these processes (Ancevska-Taneva et al. 2006; Chapple and Cheetham 2003; Lanctôt et al. 2006; Neuhaus et al. 2006).

These receptors are of particular relevance to human health. Since most hormones transmit a signal through GPCRs, many diseases associated with mutations in the GPCRs disrupt the endocrine system. In these diseases there is usually a violation of receptor trafficking to the surface of the plasma membrane (Conn et al. 2007). This happens because in the endoplasmic reticulum the resident HSP70 Bip recognizes the mutant GPCR as a mis-folded protein and sends this GPCR for proteasomal degradation (Conn et al. 2007; Anelli and Sitia 2008).

2.4.1 What New we did Learn from this Chapter (Summary) and P.S.

We have examined only the most important part of what is known about HSP70 functions in normal cells. We have learnt that:

1. The HSP70 ATPase cycle together with co-chaperones forms the protein folding mechanism and provides protein transportation into organelles.
2. The HSP70 ATPase cycle assisted by HSP40 recognizes irreversibly damaged proteins and, assisted by CHIP, Bag-1 and HSJ1 ubiquitinates these proteins and sends them for proteasomal degradation.
3. The HSP70/HSP90 complex, together with the co-chaperones Hsp40, Hip, Hop and Bag-1, recognizes normal proteins containing the marker sequence KFPRQ and sends these proteins to the lysosome for degradation.
4. The HSP70 ATPase cycle forms the protein quality control system (FORD machinery) and, depending on the state of the protein, sends the protein either for re-folding or for degradation. Thus, the FORD machinery maintains protein homeostasis.
5. Transitions between the "folding" and "degradation" states of the cycle are carefully monitored by HSP90 and by the ratio of co-chaperones, folding stimulators and degradation stimulators.

6. The HSP70 ATPase cycle controls the activity of key signalling proteins by keeping them in an inactive or active state, by regulation of their levels and by intracellular transport.

P.S. The multitude of HSP70 functions and cellular processes is astounding! This protein pokes its nose everywhere! Now we understand that there is safety in numbers and that the HSP70 multifunctional uniqueness is connected both with its three-domain structure and, primarily, with a large number of its co-chaperone "assistants". First HSP70 attracts co-chaperones, and then, with their help, it builds a multifunctional regulatory network that is used by the cell for many purposes: folding, re-folding, degradation or regulation of signalling transduction events. Truly, first the "the king makes the suite" and then "the suite makes the king"!

Despite the significant progress in understanding how this network works, many questions still remain unanswered. We do not understand how the co-chaperones direct HSP70 to substrates; how HSP70 knows that this particular protein is to be folded, but the other one should be sent for degradation; what determines the substrate specificity of HSP70 to signalling proteins? There are many more other questions that remain unanswered.

Therefore, many more discoveries about cell processes that depend on the activity of HSP70 will undoubtedly be found in the future. In relation to the study of HSP70 effects it would be appropriate now to recall the words of the great British Prime Minister Winston Churchill, "Now this is not the end. It is not even the beginning of the end. But it is, perhaps, the end of the beginning" (Fig. 2.7).

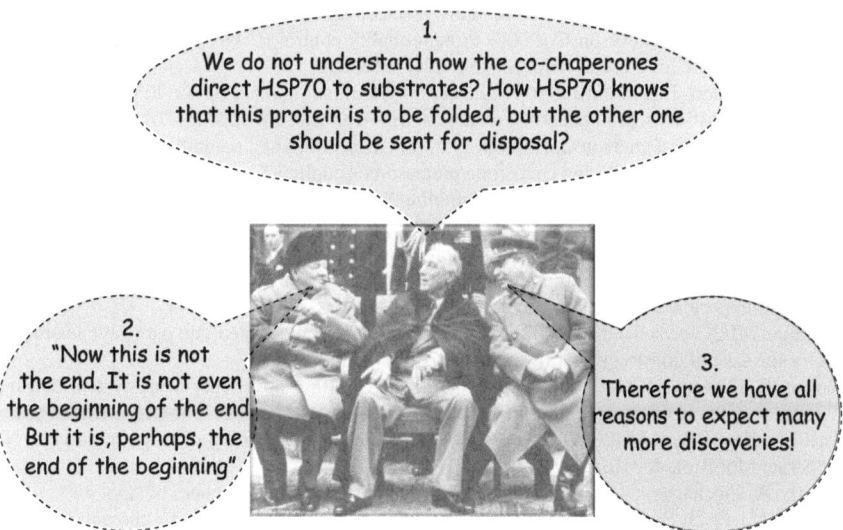

Fig. 2.7 Discussion of the important problems

References

Alberti S, Esser C, Höhfeld J (2003) BAG-1–a nucleotide exchange factor of Hsc70 with multiple cellular functions. Cell Stress Chaperon 8(3):225–231

Alberti S, Böhse K, Arndt V et al (2004) The cochaperone HspBP1 inhibits the CHIP ubiquitin ligase and stimulates the maturation of the cystic fibrosis transmembrane conductance regulator. Mol Biol Cell 15(9):4003–4010

Ancevska-Taneva N, Onoprishvili I, Andria ML et al (2006) A member of the heat shock protein 40 family, hlj1, binds to the carboxyl tail of the human mu opioid receptor. Brain Res 1081(1):28–33

Anelli T, Sitia R (2008) Protein quality control in the early secretory pathway. EMBO J 27(2):315–327

Anfinsen CB (1973) Principles that govern the folding of protein chains. Science 181:223–230

Arndt V, Daniel C, Nastainczyk W et al (2005) BAG-2 acts as an inhibitor of the chaperone-associated ubiquitin ligase CHIP. Mol Biol Cell 16(12):5891–5900

Ballinger CA, Connell P, Wu Y et al (1999) Identification of CHIP, a novel tetratricopeptide repeat-containing protein that interacts with heat shock proteins and negatively regulates chaperone functions. Mol Cell Biol 19(6):4535–4545

Bercovich B, Stancovski I, Mayer A et al (1997) Ubiquitin-dependent degradation of certain protein substrates in vitro requires the molecular chaperone Hsc70. J Biol Chem 272(14):9002–9010

Bukau B, Weissman J, Horwich A (2006) Molecular chaperones and protein quality control. Cell 125(3):443–451

Chapple JP, Cheetham ME (2003) The chaperone environment at the cytoplasmic face of the endoplasmic reticulum can modulate rhodopsin processing and inclusion formation. J Biol Chem 278(21):19087–19094

Chiang HL, Terlecky SR, Plant CP, Dice JF (1989) A role for a 70-kilodalton heat shock protein in lysosomal degradation of intracellular proteins. Science 246(4928):382–385

Conn PM, Ulloa-Aguirre A, Ito J, Janovick JA (2007) G protein-coupled receptor trafficking in health and disease: lessons learned to prepare for therapeutic mutant rescue in vivo. Pharmacol Rev 59(3):225–250

Connell P, Ballinger CA, Jiang J et al (2001) The co-chaperone CHIP regulates protein triage decisions mediated by heat-shock proteins. Nat Cell Biol 3(1):93–96

Cyr DM, Höhfeld J, Patterson C (2002) Protein quality control: U-box-containing E3 ubiquitin ligases join the fold. Trends Biochem Sci 27(7):368–375

Demand J, Lüders J, Höhfeld J (1998) The carboxy-terminal domain of Hsc70 provides binding sites for a distinct set of chaperone cofactors. Mol Cell Biol 18(4):2023–2028

Demand J, Alberti S, Patterson C, Höhfeld J (2001) Cooperation of a ubiquitin domain protein and an E3 ubiquitin ligase during chaperone/proteasome coupling. Curr Biol 11(20):1569–1577

Dobson CM (2003) Protein folding and misfolding. Nature 426(6968):884–890

Elliott E, Tsvetkov P, Ginzburg I (2007) BAG-1 associates with Hsc70. Tau complex and regulates the proteasomal degradation of Tau protein. J Biol Chem 282(51):37276–37284

Esser C, Alberti S, Höhfeld J (2004) Cooperation of molecular chaperones with the ubiquitin/proteasome system. Biochim Biophys Acta 1695(1–3):171–188

Glickman MH, Ciechanover A (2002) The ubiquitin-proteasome proteolytic pathway: destruction for the sake of construction. Physiol Rev 82(2):373–428

Glover JR, Lindquist S (1998) Hsp104, Hsp70, and Hsp40: a novel chaperone system that rescues previously aggregated proteins. Cell 94(1):73–82

Hartl FU, Hayer-Hartl M (2009) Converging concepts of protein folding in vitro and in vivo. Nat Struct Mol Biol. doi:10.1038/nsmb.1591

Hershko A, Ciechanover A (1998) The ubiquitin system. Annu Rev Biochem 67:425–479

Höhfeld J, Jentsch S (1997) GrpE-like regulation of the hsc70 chaperone by the anti-apoptotic protein BAG-1. EMBO J 16(20):6209–6216

Jäättelä M (1999) Escaping cell death: survival proteins in cancer. Exp Cell Res 248(1):30–43

Jiang J, Ballinger CA, Wu Y et al (2001) CHIP is a U-box-dependent E3 ubiquitin ligase: identification of Hsc70 as a target for ubiquitylation. J Biol Chem 276(46):42938–42944

Kregel KC (2002) Heat shock proteins: modifying factors in physiological stress responses and acquired thermotolerance. J Appl Physiol 92(5):2177–2186

Lanctôt PM, Leclerc PC, Escher E et al (2006) Role of N-glycan-dependent quality control in the cell-surface expression of the AT1 receptor. Biochem Biophys Res Commun 340(2):395–402

Lee DH, Sherman MY, Goldberg AL (1996) Involvement of the molecular chaperone Ydj1 in the ubiquitin-dependent degradation of short-lived and abnormal proteins in *Saccharomyces cerevisiae*. Mol Cell Biol 16(9):4773–4781

Lüders J, Demand J, Höhfeld J (2006) The ubiquitin-related BAG-1 provides a link between the molecular chaperones Hsc70/Hsp70 and the proteasome. J Biol Chem 275(7):4613–4617

Luo W, Zhong J, Chang R et al (2010) Hsp70 and CHIP selectively mediate ubiquitination and degradation of hypoxia-inducible factor (HIF)-1alpha but Not HIF-2alpha. J Biol Chem 285(6):3651–3663

Majeski AE, Dice JF (2004) Mechanisms of chaperone-mediated autophagy. Int J Biochem Cell Biol 36(12):2435–2444

Meacham GC, Patterson C, Zhang W et al (2001) The Hsc70 co-chaperone CHIP targets immature CFTR for proteasomal degradation. Nat Cell Biol 3(1):100–105

Molinari M, Galli C, Piccaluga V et al (2002) Sequential assistance of molecular chaperones and transient formation of covalent complexes during protein degradation from the ER. J Cell Biol 158(2):247–257

Neuhaus EM, Mashukova A, Zhang W et al (2006) A specific heat shock protein enhances the expression of mammalian olfactory receptor proteins. Chem Senses 31(5):445–452

Nishikawa S, Brodsky JL, Nakatsukasa K (2005) Roles of molecular chaperones in endoplasmic reticulum (ER) quality control and ER-associated degradation (ERAD). J Biochem 137(5):551–555

Petrucelli L, Dickson D, Kehoe K et al (2004) CHIP and Hsp70 regulate tau ubiquitination, degradation and aggregation. Hum Mol Genet 13(7):703–714

Pratt WB (1997) The role of the hsp90-based chaperone system in signal transduction by nuclear receptors and receptors signaling via MAP kinase. Annu Rev Pharmacol Toxicol 37:297–326

Pratt WB, Toft DO (2003) Regulation of signaling protein function and trafficking by the hsp90/hsp70-based chaperone machinery. Exp Biol Med (Maywood) 228(2):111–133

Salvador N, Aguado C, Horst M, Knecht E (2000) Import of a cytosolic protein into lysosomes by chaperone-mediated autophagy depends on its folding state. J Biol Chem 275(35):27447–27456

Sepp-Lorenzino L, Ma Z, Lebwohl DE et al (1995) Herbimycin A induces the 20 S proteasome- and ubiquitin-dependent degradation of receptor tyrosine kinases. J Biol Chem 270(28):16580–16587

Westhoff B, Chapple JP, van der Spuy J et al (2005) HSJ1 is a neuronal shuttling factor for the sorting of chaperone clients to the proteasome. Curr Biol 15(11):1058–1064

Whitesell L, Mimnaugh EG, De Costa B et al (1994) Inhibition of heat shock protein HSP90-pp 60v-src heteroprotein complex formation by benzoquinone ansamycins: essential role for stress proteins in oncogenic transformation. Proc Natl Acad Sci USA 91(18):8324–8328

Zhang Y, Nijbroek G, Sullivan ML et al (2001) Hsp70 molecular chaperone facilitates endoplasmic reticulum-associated protein degradation of cystic fibrosis transmembrane conductance regulator in yeast. Mol Biol Cell 12(5):1303–1314

Chapter 3
HSP70 in Damaged Cells

Abstract In a damaged cell HSP70 maintains protein homeostasis. To achieve this, HSP70, together with co-chaperones, prevents protein aggregation, aids in the dissociation of formed protein aggregates, and targets particular "irreparable" proteins for degradation. In addition, because of HIF-1 activation, the restoration of protein homeostasis forms a specific cell defense against hypoxic injury, against free-radical injury owing to an increase in antioxidant activity, and against calcium injury owing to a reduction in the calcium level in the cell. HSP70 can deposit mutant proteins. However, such mutant proteins can be released when denatured proteins appear in the cell. HSP70 blocks apoptosis by inhibiting the release of proapoptotic factors from mitochondria, inhibiting AIF, caspase-9 and JNK activities, as well as by increasing the Bcl-2 level and decreasing the Bax level. HSP70 protects cells from the accidental triggering of apoptosis by restricting DNA-ase folding until the inhibitor binds to this proapoptotic protein. All of the above processes lead to the general conclusion: *HSP70 is a component of an intracellular system aimed at maintaining protein homeostasis and protecting damaged cells.*

Keywords HSP70 • Protein homeostasis • Cell damage • Cell defence • Apoptosis

In the previous chapter we looked at the functions of HSP70 in normal cells and learned that: (i) the HSP70 ATPase cycle forms a protein quality control system (FORD machinery) that, depending on the protein condition, sends the protein either for folding/re-folding or for degradation. Consequently, the FORD machinery maintains protein homeostasis. (ii) The HSP70 ATPase cycle controls the activity of key signaling proteins. These proteins include the steroid hormone receptors, Raf kinase, eIF2α-kinase and CyclinB1/Cdk1, and transcription factors HSF-1, c-Myc and pRb.

Studies of HSP70 functions in damaged cells have been conducted with equal intensity and this is what will be covered in this chapter. To understand better what essentially happens in a damaged cell, and to assess the role of HSP70 in these events, we need to recall the notion of homeostasis.

I. Malyshev, *Immunity, Tumors and Aging: The Role of HSP70,*
SpringerBriefs in Biochemistry and Molecular Biology,
DOI: 10.1007/978-94-007-5943-5_3, © The Author(s) 2013

3.1 The Concept of Homeostasis: From Hippocrates to the Present Day, or What Would Have Happened had Friedrich Engels and Claude Bernard Become Friends

The concept of homeostasis as a certain constancy of the internal environment was first suggested by the famous Greek physician Hippocrates. He believed that a human stays healthy for as long as the ratio of the four liquids—red from blood, yellow from liver, black from spleen and blue from brain—remain constant in the body. Many years later the physiologists Claude Bernard and Walter Cannon conferred a scientific form to these semi-mystical ideas of Hippocrates; they developed the concept of homeostasis, which until now remains as important as the Darwin theory of evolution and the Schleiden and Schwann cell theory. The very essence of the homeostasis theory was best expressed by Bernard (1878): "The constancy of the internal environment is a guarantee of free and independent life!"

At the same time in the nineteenth century, when Claude Bernard was developing his concept, Friedrich Engels, a friend and sponsor of Karl Marx, defined life as a mode of existence of protein bodies (Engels 1987). This statement is not far from truth! If Engels was a friend of Claude Bernard rather than of Karl Marx, the two of them would have certainly developed a joint formula "Homeostasis of proteins is a key condition for the existence of life". Consequently, whoever explains the mechanisms of cellular protein turnover, i.e., the mechanisms of both the synthesis of new proteins and of the removal of "old" and damaged proteins, will basically answer the question "what is life?" Let's be ambitious and try to do it here and now! After the first two chapters about the heat shock proteins we are ready for it!

3.2 HSP70 as a "Stem" Molecule of Protein Homeostasis

In the second half of the last century, scientists obtained convincing data confirming that HSP70 plays an important role in the folding of newly created proteins, in the degradation of the old and damaged ones, as well as in ensuring the normal functioning of proteins. Understanding the HSP70 ATPase cycle was very useful in figuring out how it works. Basically we are ready to conclude that the HSP70 ATPase cycle, or, FORD machinery, is a homeostatic mechanism that regulates intracellular protein turnover. However, if it is so, then the FORD machinery should follow the homeostasis principles, which are simple and well-known.

The modern concept of homeostasis has taught us that *the constant internal environment is supported by a negative feedback mechanism which consists of three elements: sensor—the structure that notes disorders of homeostasis; regulator—the structure that regulates the body or cell response to this disorder; and effector—the structure, which eliminates this disorder.*

Now we can make a small discovery: the protein quality control system, or the FORD mechanism, exactly matches the structure of a homeostatic mechanism (Fig. 3.1). The HSP40 co-chaperone and the HSP70 substrate domain, which recognize the quality of proteins, form a two-component sensor. The regulator role is played by the ATPase N-domain complex with co-chaperones, which regulate ATP hydrolysis, nucleotide exchange, as well as co-chaperone-switches of the FORD functions. Finally, the intracellular executive structures that provide folding or degradation of proteins are the effectors for the FORD machinery. Further, if a violation of cell homeostasis is related with increased synthesis of new proteins, the FORD machinery will increase the folding rate, and that will consequently reduce the the number of unfolded polypeptides and eliminate the disruption of homeostasis. If the disruption of homeostasis is associated with the appearance of old or mutant proteins, the FORD machinery will direct them to proteasomes or lysosomes for degradation and will thus restore homeostasis.

HSP70 is admirable and truly unique! A three-domain structure of one molecule contains the possibility of forming the whole three-component homeostatic mechanism of negative feedback! It would be fair to say that HSP70 is the main "stem" molecule for the whole homeostasis of cell proteins.

The notion of the homeostatic principle for the FORD operation clearly indicates that HSP70 activity in a damaged cell will be primarily directed for restoring the disrupted protein homeostasis. How HSP70 carries this process out is dependent on the nature of the disruption of protein homeostasis in a cell.

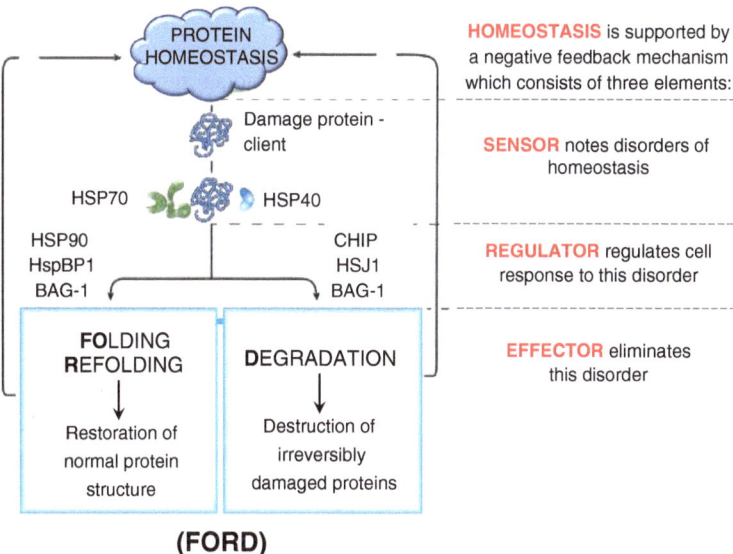

(FORD)

Fig. 3.1 The protein quality control system (FORD mechanism) exactly matches the structure of a homeostatic mechanism

3.3 The Convergence of Different Mechanisms of Cell Damage Towards the Disruption of Protein Homeostasis Turns HSP70 into a Universal Protective Factor

Any textbook on cell pathophysiology will describe three main types of cell damage: (1) hypoxic damage, (2) free-radical damage and (3) intracellular calcium overload. Even by simply skimming through the respective parts of the textbook, one would notice something important: a serious disruption of protein homeostasis is common for all these three cases with an increase in the occurrence of denatured and damaged proteins. This may be associated with developing acidosis, action of free radicals and/or activation of calcium-dependent proteases.

The loss of native conformation leads not only to the loss of functional properties of the protein, but also to another serious problem, namely polypeptide aggregation with the formation of toxic aggregates. Fortunately the appearance of denatured proteins is also a signal to activate synthesis of protective HSPs.

Approximately 25 years ago, Hugh Pelham, describing the role of HSP70 in a damaged nucleolus, was the first to hint at how HSP70 could participate in renaturation of proteins and disaggregation of protein aggregates. The Pelham hypothesis was based on the fact that HSP70 has a unique high affinity to denatured proteins. Consequently, HSP70 would readily bind to these proteins and, using the energy from ATP hydrolysis, break non-covalent hydrophobic bonds between damaged proteins and disrupt protein aggregates. As a result, the released proteins can renature to restore native structure and function (Fig. 1.3).

This hypothesis was largely confirmed; it even turned into a concept, with a number of important details added.

So how, in the modern view, does protein homeostatis recover after disruption induced by the appearance of denatured, damaged and aggregated proteins?

To maintain protein homeostasis in a damaged cell, the FORD protein quality control system uses several approaches. First, HSP70 can prevent the formation of aggregates. Second, if the aggregates have already formed, HSP70 can disaggregate them and restore protein function (Liberek et al. 2008). Finally, if the protein cannot be restored, FORD targets the protein for degradation.

HSP70 can prevent aggregation virtually alone, by binding to hydrophobic areas of denatured proteins to prevent formation of non-covalent hydrophobic interactions between proteins (Fig. 1.3). HSP70 may need HSP40 to recognize a damaged protein, and the energy from ATP to capture tightly and shield the hydrophobic region.

Additional partners, primarily the chaperone HSP100 and small HSPs, are required for disaggregating the aggregates formed (Liberek et al. 2008). The disaggregation mechanism is presented in Fig. 3.2. It is assumed that first HSP70 extracts a polypeptide from the aggregate (Ziętkiewicz et al. 2006) and then inserts this peptide into the channel of another chaperone, HSP100 (Haslberger et al. 2007). Inside this channel there are special loops that pull the polypeptides outside through the channel using ATP as an energy source (Hinnerwisch et al. 2005).

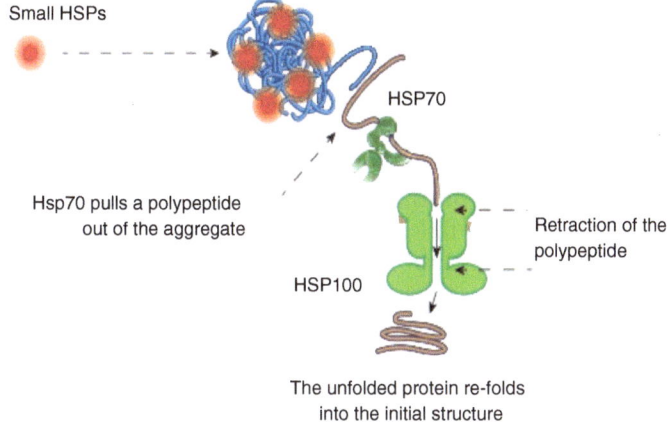

Fig. 3.2 The chaperone HSP100 and small HSPs play an important role in the "aggregation-disaggregation" processes

Retraction of the polypeptide through the channel leads to the full unfolding of the protein into a linear chain. (Lum et al. 2004; Schlieker et al. 2004; Weibezahn et al. 2004). After that, the unfolded protein re-folds into the native structure either without assistance, or with the help of the HSP70 chaperone system.

Researchers from the Dr. Walter laboratory revealed important details of this process. They discovered that, in the absence of HSP70, the polypeptide advancement through the HSP100 channel is significantly reduced (Doyle et al. 2007a, 2007b; Schaupp et al. 2007). The researchers interpreted these findings and made a logical conclusion; HSP70 not only contributes to the substrate transportation to the HSP100 channel, but also stimulates ATP hydrolysis in the HSP100 ATPase domains.

This is a very familiar situation to the process in which HSP40 transports substrates to HSP70 and stimulates ATP cleavage in the HSP70 N-domain. We can continue with this analogy. In both the HSP40-HSP70 and HSP70-HSP100 pairs, polypeptide binding to HSP70 or to HSP100 significantly stimulates ATP hydrolysis in the ATPase domain of that HSP (Schaupp et al. 2007).

In the literature there is no explanation as to why two similar mechanisms are required. One can only assume that the HSP40-HSP70 bi-chaperone cooperation is used for the folding of polypeptides synthesized on ribosomes, and that the HSP70-HSP100 complex is used for pulling polypeptides out of an aggregate and their subsequent refolding.

Another important observation describing dissociation of the protein aggregates was made by two Japanese scientists, Kitagawa and Nakamoto (Kitagawa et al. 2000; Nakamoto et al. 2000). They suggested that small HSPs (sHSPs) play an important role in controlling the "aggregation-disaggregation" processes (Fig. 3.2). Members of this family have a low molecular weight, 15–43 kDa (Haslbeck 2002; Haslbeck et al. 2005) and do not require ATP hydrolysis for manifesting their activity, i.e., they are ATP-independent.

sHSPs in damaged cells bind to denatured proteins during the process of aggregate formation (Franzmann et al. 2005; Haslbeck et al. 1999), thereby reducing the number of hydrophobic contacts between the denatured polypeptides. This substantially increases the efficiency of disaggregation and subsequent re-folding (Lee and Vierling 2000; Mogk et al. 2003; Matuszewska et al. 2005).

Unfortunately, despite all the cooperative efforts of the chaperones, not all proteins restore their native structure after disaggregation. Some of them remain damaged. There would be little "debate" with such "incorrigible" proteins: the FORD protein quality control system will send them for degradation using the mechanism that we have examined in the last chapter. Basically the HSP70 ATPase cycle, assisted by the co-chaperones, will recognize the irreversibly damaged proteins and, with the help of CHIP, Bag-1 and HSJ1, will ubiquitinate them and send them to proteasomes for degradation.

So, irrespective of the type of cell damage (hypoxic, radical or calcium) HSP70 contributes to cell recovery through restoration of the impaired protein homeostasis. To achieve this "strategic" goal, HSP70 and its partners employs three "tactics": *preventive*—HSP70 shields hydrophobic areas of damaged proteins and prevents their aggregation; *therapeutic*—HSP70 disaggregates protein formations and facilitates refolding of the disaggregated proteins; and finally the *surgical* tactic—HSP70 labels irreversibly damaged proteins for degradation.

We have just examined how HSP70 can assist in restoring protein homestasis, disturbed during various types of cell damage. This is precisely a situation where many would say that, since the effect is evident under a variety of damaging conditions, this effect is non-specific. It was noticed a long time ago that, as soon as scientists fail to study the problem in depth, they resort to the notion of "non-specificity". In our case, the seemingly "non-specific" restoration of protein homeostasis actually has very "specific" protective consequences.

For instance, as mentioned before, normally, HSP70 and CHIP interact with the transcription factor HIF-1. Such an interaction leads to HIF-1 ubiquitination and subsequent removal via proteasomal degradation (Luo et al. 2010). As a result, HIF-dependent genes remain inactive. Under hypoxia, denatured proteins appear in the cell. This leads to detachment of HSP70 from HIF-1 and the binding to damaged proteins to which HSP70 has higher affinity. As a result, HSP70 no longer sends HIF-1 for degradation and the active HIF-1 can enter the nucleus to activate genes responsible for adaptation to hypoxia, such as the glycolytic enzymes.

It was shown also that, in the case of free-radical damage, HSP70 can limit free radical generation by damaged mitochondria (Ouyang et al. 2006; Xu and Giffard 1997) and stimulate the activity of antioxidant enzymes, such as catalase, glutathione peroxidase (Romero et al. 2010) and superoxide dismutase (Suzuki et al. 2002).

Finally, in the case of intracellular calcium overload, it was found that an increase in HSP70 levels correlates with a decrease in calcium levels in the cell (Szenczi et al. 2005; Sharma et al. 2003). These examples represent specific cases where HSP70 plays a role in protecting the cell.

Protein homeostasis can be disrupted without a cell being damaged, for instance when mutant proteins appear. This is another interesting story, in which HSP70 plays an important role.

Mutants of p53,[1] CFTR or superoxide dismutase (SOD) draw the particular attention of HSP70 (Gaiddon et al. 2001; Shinder et al. 2001). HSP70 strongly binds to such proteins and does not release them, thus preventing these mutants from manifesting their negative effects (Fig. 3.3). This is clearly a good function of HSP70, but at the same time it becomes a true time bomb! Such mutant proteins can be released when denatured proteins appear in the cell. The denatured proteins will attract HSP70 molecules that are in complex with the mutants. It can, for instance, happen under stress conditions, at some stages of development or aging, when HSP70 synthesis decreases (Neupert and Brunner 2002; Rogue et al. 1993). These mechanisms may contribute to the development of pathologies, such as oncogenesis, when a mutant p53 is released, neurodegenerative diseases such as

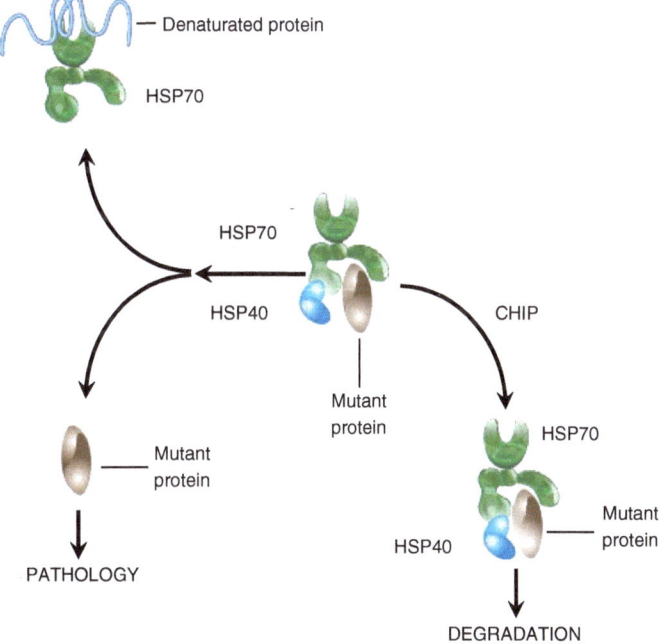

Fig. 3.3 HSP70 can deposit mutant proteins and can release such mutant proteins when denatured proteins appear in the cell

---

[1] p53—a protein product of a tumor suppressor gene, regulates cell growth and proliferation, and prevents unrestrained cell division after chromosomal damage, as from ultraviolet or ionizing radiation.The absence of p53 as a result of a gene mutation increases the risk of developing various cancers.

amyotrophic lateral sclerosis when a mutant SOD is released, Parkinsonism when a mutant α-synuclein is released and Huntington's chorea when a mutant huntingtin protein is released. Alternatively, HSP70 and its co-chaperones can degrade mutant proteins. Such events were shown in relation to mutant CFTR and mutant SOD (Meacham et al. 2001; Urushitani et al. 2004).

3.4 HSP70 and Apoptosis

Finally, in all cases of cell damage, or when mutant proteins appear, in the most severe situation the cell dies via an apoptotic cell-death type process. Apoptosis can also be induced by activation of death receptors, like Fas or tumor necrosis factor (TNF) receptors, decreased levels of growth factors, excessive DNA damage, and the effects of drugs and radiation (Martin et al. 1995; Rosette and Karin 1995).

Before inducing apoptosis in a cell culture researchers at one Californian laboratory activated HSP70 synthesis in cells by heat treatment. The result was astounding! In the control group apoptosis developed and the cells died. In contrast, apoptosis did not develop in the cells with a high concentration of HSP70. With this in view one starts respecting the Finnish sauna even more; as researchers (naturally Finnish) have shown, the sauna does activate the synthesis of protective HSP70. Thus it was shown that HSP70 can exhibit antiapoptotic properties.

Originally it was proposed that the antiapoptotic effect of HSP70 was associated with the ability of HSP70 to support protein folding and to limit aggregation. Subsequent research showed; however, that HSP70 can directly interfere with the apoptotic program (Giffard and Yenari 2004; Papadopoulos MC et al. 1996; Sun Y et al. 2006).

Apoptosis can be triggered by two main pathways (Leist and Jäättelä 2001): mitochondria-dependent and receptor-dependent (Fig. 3.4).

Mitochondria-dependent apoptosis can be activated in response to cell stress and intracellular changes induced, for instance, by ischemia (Chan 2004; Matsumori et al. 2006; Matsumoto et al. 1999). In this case, apoptosis is initiated by release from mitochondria of proapoptotic molecules, such as cytochrome c, AIF (apoptosis inducing factor), Smac/DIABLO (Gogvadze and Orrenius 2006) and endonuclease G (EndoG) (Garrido and Kroemer 2004). In the cytoplasm, cytochrome c interacts with Apaf-1 (apoptosis protease activating factor-1) and dATP to form an apoptosome that activates specific protease caspase-9 (Gogvadze and Orrenius 2006; Leist and Jäättelä 2001). Caspase-9 then triggers the activation mechanisms of other caspases (Slee et al. 1999).

The triggering of mitochondria-dependent apoptosis is closely monitored by the Bcl-2 protein family. Bcl-2 is a key antiapoptic member of the family. It blocks the release of cytochrome c and AIF from mitochondria, and thus prevents the activation of caspases. (Merry and Korsmeyer 1997; Yuan and Yankner 2000).

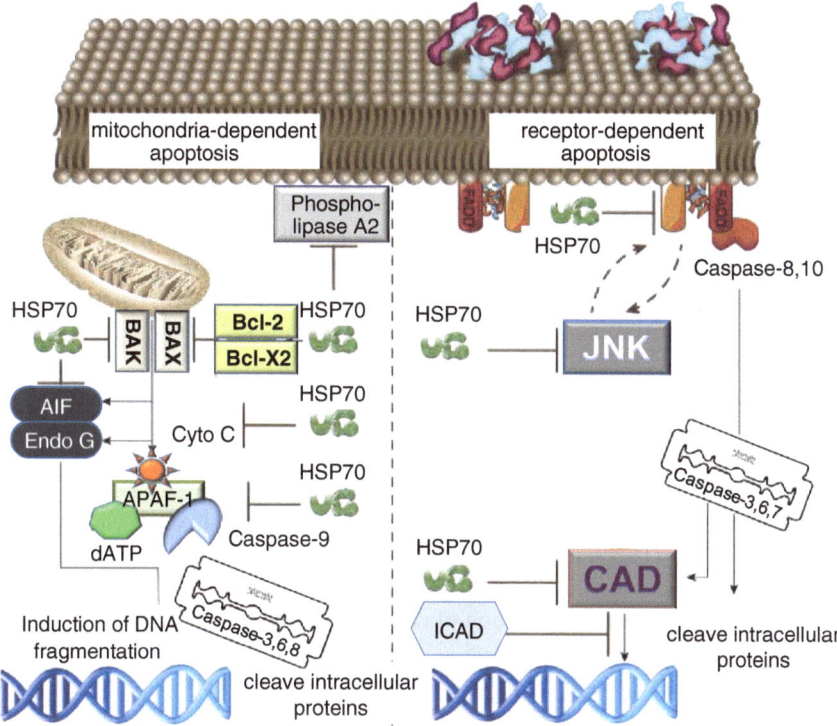

Fig. 3.4 The mechanisms of antiapoptotic effects of HSP70

The Bax protein, a counter-partner to Bcl-2, acts as a proapoptic factor. The Bcl-2/Bax balance determines whether a cell will take the apoptotic pathway or not.

Receptor-dependent apoptosis is triggered by death ligands, such as Fas, TNF-α and the death receptor ligand (DRL), which activate death receptors on the plasma membrane. Through their death domains these receptors activate caspase-8, which in turn activates caspase-3 (Thorburn 2004).

The increase in the number of death receptors may be due to the activation of the protein kinase c-Jun N-terminal kinase (JNK). The activation of JNK itself is stimulated by death receptors (Kitamura et al. 2003).

In both mitochondrial-dependent and receptor-dependent apoptosis, activated caspases cleave intracellular proteins, thereby reverting apoptosis to an irreversible event (Andrabi et al. 2006; Kauppinen and Swanson 2007). In addition, caspase-3 activates caspase-activated DNAase that fragments DNA.

Apoptosis may develop even without caspase activation, due to translocation of AIF and endonuclease G (EndoG) to the nucleus and induction of chromatin condensation and DNA fragmentation.

HSP70 has been shown to block the development of the apoptotic program through several mechanisms (Ravagnan et al. 2001; Stankiewicz et al. 2005; Steel et al. 2004).

At the mitochondrial level. It was shown that an increase in the HSP70 level leads to an increase in antiapoptotic protein Bcl-2 (Kelly et al. 2002) and a decrease in proapoptotic Bax (Stankiewicz et al. 2005). HSP70 blocks the possibility for Bax to incorporate into the external mitochondrial membrane, thus preventing the increase in permeability of mitochondrial membranes and the exit of cytochrome c and AIF (Stankiewicz et al. 2005). In addition, HSP70 can directly block the exit of cytochrome c (Lee et al. 2004; Matsumori et al. 2006; Tsuchiya et al. 2003), the Smac/DIABLO protoapoptic protein (Jiang et al. 2005) and Apaf-1(Beere et al. 2000; Matsumori et al. 2006; Saleh et al. 2000) from mitochondria.

At the post-mitochondrial level. HSP70 can bind to Apaf-1 thus preventing the attraction of procaspase-9 to the apoptosome (Saleh et al. 2000), and this complex also inhibits caspase-9 activity (Beere et al. 2000).

At the level of death receptors. HSP70 can reduce the number of death receptors because of inhibition of JNK (Lee et al. 2005; Park et al. 2001). In addition, HSP70 can bind to death receptors DR4 and DR5 to inhibit the passage of apoptotic signals through these receptors (Guo et al. 2005).

At the final stages of apoptosis. HSP70 can limit apoptosis when caspase activation has already happened; for instance, it can restrict activation of phospholipase A2 and changes in nuclear morphology (Jaattela et al. 1998).

HSP70 can also prevent activation of **caspase-independent** apoptotic pathways (Creagh et al. 2000; Ravagnan et al. 2001) because of the interaction with AIF and EndoG, and thus block the translocation of these proapoptotic factors into the nucleus (Gurbuxani et al. 2003; Kalinowska et al. 2005; Matsumori et al. 2005; Ravagnan et al. 2001; Ruchalski et al. 2006; Sun et al. 2006).

In general, apoptosis is a very serious process. It is needed, first of all, to eliminate from the body cells with dangerous point mutations, cells that do not perform their functions any more, and cells that are old or damaged. For this purpose each cell already has all necessary molecular instruments for its own "suicide": caspases and DNAases. It is like a story of one ancient tribe in the Amasonian forest. Chiefs of that tribe had a very beautiful sword hidden in a particular place. Whenever a chief made a mistake, at war or while hunting, he would take the sword out and, according to that tribe's law, would kill himself! Japanese samurai followed a similar code of conduct.

However, while no mistake has been made, there is no need to kill yourself. That is why the caspase "swords" in a cell are—until the time comes—inactive in the form of procaspases, and DNAases are bound to their inhibitors. In this way, nature made absolutely sure that apoptosis is not activated by accident. However, in the case of caspase-activated DNAase (CAD) the cell "story" of apoptosis has one subtlety. While de novo synthesis takes place, DNAase is not bound to its inhibitor; so immediately after the translation is over, and before the inhibitor binds, it may cause a lot of trouble in the cell.

To solve this problem the cell uses HSP70. HSP70 binds CAD during the translation process in ribosomes, preventing the enzyme from folding. The folding is completed only when the CAD inhibitor ICAD is added to CAD (Sakahira and Nagata 2002).

Therefore, HSP70 not only inhibits the activation of apoptosis, but also prohibits an accidental triggering of apoptosis when dangerous proapoptic factors are synthetized de novo.

Such an important role of HSP70 in controlling the development of apoptosis means that disruption of HSP70 synthesis or functioning would lead to the development of serious diseases. Many tumor cells, for instance, show increased levels of HSP70; this correlates with an enhanced resistance of tumor cells to apoptosis and with tumor growth (Nylandsted et al. 2000). In contrast, neurodegenerative diseases such as Alzheimer's disease, Parkinsonism, Huntington chorea and amyotrophic lateral sclerosis are associated with a decrease in the synthesis of HSP70 and therefore excessive apoptosis.

When we are analyzing the protective role of HSP70 in a damaged cell, we need to take into account another critically important point. As a rule, in all cases of cell damage ATP synthesis is disrupted and an energy shortage arises. Here we need to recall that homeostatic mechanisms of the FORD machinery, as well as the effects of HSP70, are ATP-dependent. So it is important to remember that HSP70 can protect the cell only when a decrease in the ATP level is not critical for the functioning of the HSP70 ATPase cycle. Whenever the ATP level dramatically decreases (and this may happen with any type of damage) neither FORD nor HSP70 will be able to perform their functions. As a result, HSP70 will no longer function as a homeostatic regulator, but rather a pathogenetic part of irreversible cell damage.

3.5 What New Have we Learnt From This Chapter (Summary)

In this chapter the dramatic story of HSP70 in a damaged cell was presented and we can summarize what we have learnt.

1. The protein quality control system (FORD-machinery) matches the structure of the homeostatic mechanism of negative feedback.

2. To maintain protein homeostasis in a damaged cell, the FORD-machinery prevents protein aggregation, disrupts formed aggregates, and labels "irreparable" proteins for proteasomal degradation.

3. In the initial stages of the protein disaggregation process, HSP70 extracts a protein from the aggregate and inserts it into the HSP100 channel. Pulling the protein through the channel leads to the protein unfolding and refolding to the native structure.

4. During the process of aggregate formation sHSPs bind to denatured proteins. This reduces the number of hydrophobic contacts and substantially increases the efficiency of protein disaggregation by other HSPs such as HSP70.

5. HSP70 forms a specific cell defense: (1) against hypoxic injury due to HIF-1; (2) against free-radical injury due to increases in antioxidant activity; and (3) against calcium injury due to a decrease in the calcium levels of the cell.

6. HSP70 can deposit mutant proteins. However, such mutant proteins can be released when denatured proteins appear in the cell.

7. HSP70 blocks apoptosis by inhibiting the release of proapoptotic factors from mitochondria, inhibits proteins such as AIF, caspase-9 and JNK, as well as regulate an increase in the Bcl-2 level and a decrease in the Bax level.

8. HSP70 guarantees against unwanted apoptotic events by inhibiting the activity of newly produced DNAase until the inhibitor protein binds to this proapoptotic protein.

All of the above allows us to make a general conclusion: *HSP70 is a component of an intracellular system aimed at maintaining protein homeostasis and protecting damaged cells.*

References

Andrabi SA, Kim NS, Yu SW et al (2006) Poly(ADP-ribose) (PAR) polymer is a death signal. Proc Natl Acad Sci USA 103(48):18308–18313

Beere HM, Wolf BB, Cain K et al (2000) Heat-shock protein 70 inhibits apoptosis by preventing recruitment of procaspase-9 to the Apaf-1 apoptosome. Nat Cell Biol 2(8):469–475

Bernard C (1878) Lectures on the phenomena common to animals and plants. In: Hoff HE, Guillemin R, Guillemin L, Charles C Thomas (1974) (trans: Springfield IL)

Chan PH (2004) Mitochondria and neuronal death/survival signaling pathways in cerebral ischemia. Neurochem Res 29(11):1943–1949

Creagh EM, Carmody RJ, Cotter TG (2000) Heat shock protein 70 inhibits caspase-dependent and -independent apoptosis in Jurkat T cells. Exp Cell Res 257:58–66

Doyle SM, Shorter J, Zolkiewski M, Hoskins JR et al (2007a) Asymmetric deceleration of ClpB or Hsp104 ATPase activity unleashes protein-remodeling activity. Nat Struct Mol Biol 14(2):114–122

Doyle SM, Hoskins JR, Wickner S (2007b) Collaboration between the ClpB AAA+ remodeling protein and the DnaK chaperone system. Proc Natl Acad Sci USA 104(27):11138–11144

Engels (1987) Anti-Dühring. Marx Engels Collected Works (MECW), London

Franzmann TM, Wühr M, Richter K, Walter S, Buchner J (2005) The activation mechanism of Hsp26 does not require dissociation of the oligomer. J Mol Biol 350(5):1083–1093

Garrido C, Kroemer G (2004) Life's smile, death's grin: Vital functions of apoptosis-executing proteins. Curr Opin Cell Biol 16:639–646

Gaiddon C, Lokshin M, Ahn J, Zhang T, Prives C (2001) A subset of tumor-derived mutant forms of p53 down-regulate p63 and p73 through a direct interaction with the p53 core domain. Mol Cell Biol 21(5):1874–1887

Giffard RG, Yenari MA (2004) Many mechanisms for hsp70 protection from cerebral ischemia. J Neurosurg Anesthesiol 16(1):53–61

Gogvadze V, Orrenius S (2006) Mitochondrial regulation of apoptotic cell death. Chem Biol Interact 163(1–2):4–14

Guo F, Sigua C, Bali P et al (2005) Mechanistic role of heat shock protein 70 in Bcr-Abl-mediated resistance to apoptosis in human acute leukemia cells. Blood 105:1246–1255

Gurbuxani S, Schmitt E, Cande C et al (2003) Heat shock protein 70 binding inhibits the nuclear import of apoptosis-inducing factor. Oncogene 22:6669–6678

Haslbeck M, Walke S, Stromer T et al (1999) Hsp26: a temperature-regulated chaperone. EMBO J 18(23):6744–6751

Haslbeck M, Franzmann T, Weinfurtner D, Buchner J (2005) Some like it hot: the structure and function of small heat-shock proteins. Nat Struct Mol Biol 12(10):842–846

Haslberger T, Weibezahn J, Zahn R et al (2007) M domains couple the ClpB threading motor with the DnaK chaperone activity. Mol Cell 25(2):247–260

Haslbeck M (2002) sHsps and their role in the chaperone network. Cell Mol Life Sci 59(10):1649–1657

Hinnerwisch J, Fenton WA, Furtak KJ, Farr GW, Horwich AL (2005) Loops in the central channel of ClpA chaperone mediate protein binding, unfolding, and translocation. Cell 121(7):1029–1041

Jäättelä M, Wissing D, Kokholm K, Kallunki T, Egeblad M (1998) Hsp70 exertsitsanti-apoptotic function downstream of caspase-3-like proteases. EMBO J 17:6124–6134

Jiang B, Xiao W, Shi Y, Liu M, Xiao X (2005) Heat shock pretreatment inhibited the release of Smac/DIABLO from mitochondria and apoptosis induced by hydrogen peroxide in cardiomyocytes and C2C12 myogenic cells. Cell Stress Chaperones 10(3):252–262

Kalinowska M, Garncarz W, Pietrowska M, Garrard WT, Widlak P (2005) Regulation of the human apoptotic DNase/RNase Endonuclease G: Involvement of Hsp70 and ATP. Apoptosis 10:821–830

Kauppinen TM, Swanson RA (2007) The role of poly(ADP-ribose) polymerase-1 in CNS disease. Neuroscience 145(4):1267–1272

Kelly S, Zhang ZJ, Zhao H et al (2002) Gene transfer of HSP72 protects cornu ammonis 1 region of the hippocampus neurons from global ischemia: influence of Bcl-2. Ann Neurol 52(2):160–167

Kitagawa M, Matsumura Y, Tsuchido T (2000) Small heat shock proteins, IbpA and IbpB, are involved in resistances to heat and superoxide stresses in Escherichia coli. FEMS Microbiol Lett 184(2):165–171

Kitamura C, Ogawa Y, Nishihara T, Morotomi T, Terashita M (2003) Transient co-localization of c-Jun N-terminal kinase and c-Jun with heat shock protein 70 in pulp cells during apoptosis. J Dent Res 82(2):91–95

Lee GJ, Vierling E (2000) A small heat shock protein cooperates with heat shock protein 70 systems to reactivate a heat-denatured protein. Plant Physiol 122(1):189–198

Lee SH, Kwon HM, Kim YJ, Lee KM, Kim M, Yoon BW (2004) Effects of hsp70.1 gene knockout on the mitochondrial apoptotic pathway after focal cerebral ischemia. Stroke 35(9):2195–2199

Lee JS, Lee JJ, Seo JS (2005) HSP70 deficiency results in activation of c-Jun N-terminal Kinase, extracellular signal-regulated kinase, and caspase-3 in hyperosmolarity-induced apoptosis. J Biol Chem 280(8):6634–6641

Leist M, Jäättelä M (2001) Four deaths and a funeral: from caspases to alternative mechanisms. Nat Rev Mol Cell Biol 2(8):589–598

Liberek K, Lewandowska A, Zietkiewicz S (2008) Chaperones in control of protein disaggregation. EMBO J 27(2):328–335

Lum R, Tkach JM, Vierling E, Glover JR (2004) Evidence for an unfolding/threading mechanism for protein disaggregation by Saccharomyces cerevisiae Hsp104. J Biol Chem 279(28):29139–29146

Luo W, Zhong J, Chang R, Hu H, Pandey A, Semenza GL (2010) Hsp70 and CHIP selectively mediate ubiquitination and degradation of hypoxia-inducible factor (HIF)-1alpha but Not HIF-2alpha. J Biol Chem 285(6):3651–3653

Martin SJ, Newmeyer DD, Mathias S et al (1995) Cell-free reconstitution of Fas-, UV radiation- and ceramide-induced apoptosis. EMBO J 14(21):5191–5200

Matsumori Y, Hong SM, Aoyama K et al (2005) Hsp70 overexpression sequesters AIF and reduces neonatal hypoxic/ischemic brain injury. J Cereb Blood Flow Metab 25:899–910

Matsumori Y, Northington FJ, Hong SM et al (2006) Reduction of caspase-8 and -9 cleavage is associated with increased c-FLIP and increased binding of Apaf-1 and Hsp70 after neonatal hypoxic/ischemic injury in mice overexpressing Hsp70. Stroke 37(2):507–512

Matsumoto S, Friberg H, Ferrand-Drake M, Wieloch T (1999) Blockade of the mitochondrial permeability transition pore diminishes infarct size in the rat after transient middle cerebral artery occlusion. J Cereb Blood Flow Metab 19(7):736–741

Matuszewska M, Kuczyńska-Wiśnik D, Laskowska E, Liberek K (2005) The small heat shock protein IbpA of Escherichia coli cooperates with IbpB in stabilization of thermally aggregated proteins in a disaggregation competent state. J Biol Chem 280(13):12292–12298

Meacham GC, Patterson C, Zhang W, Younger JM, Cyr DM (2001) The Hsc70 co-chaperone
    CHIP targets immature CFTR for proteasomal degradation. Nat Cell Biol 3(1):100–105
Merry DE, Korsmeyer SJ (1997) Bcl-2 gene family in the nervous system. Annu Rev Neurosci
    20:245–267
Mogk A, Deuerling E, Vorderwülbecke S, Vierling E, Bukau B (2003) Small heat shock proteins,
    ClpB and the DnaK system form a functional triade in reversing protein aggregation. Mol
    Microbiol 50(2):585–595
Nakamoto H, Suzuki N, Roy SK (2000) Constitutive expression of a small heat-shock protein
    confers cellular thermotolerance and thermal protection to the photosynthetic apparatus in
    cyanobacteria. FEBS Lett 483(2–3):169–174
Neupert W, Brunner M (2002) The protein import motor of mitochondria. Nat Rev Mol Cell Biol
    3(8):555–565
Nylandsted J, Rohde M, Brand K, Bastholm L, Elling F, Jäättelä M (2000) Selective depletion of
    heat shock protein 70 (Hsp70) activates a tumor-specific death program that is independent
    of caspases and bypasses Bcl-2. Proc Natl Acad Sci USA 97(14):7871–7876
Ouyang YB, Xu LJ, Sun YJ, Giffard RG (2006) Overexpression of inducible heat shock protein
    70 and its mutants in astrocytes is associated with maintenance of mitochondrial physiology
    during glucose deprivation stress. Cell Stress Chaperones 11(2):180–186
Papadopoulos MC, Sun XY, Cao J, Mivechi NF, Giffard RG (1996) Over-expression of HSP-70
    protects astrocytes from combined oxygen-glucose deprivation. NeuroReport 7(2):429–432
Park HS, Lee JS, Huh SH, Seo JS, Choi EJ (2001) Hsp72 functions as a natural inhibitory protein
    of c-Jun N-terminal kinase. EMBO J 20(3):446–456
Ravagnan L, Gurbuxani S, Susin SA et al (2001) Heat-shock protein 70 antagonizes apoptosis-
    inducing factor. Nat Cell Biol 3(9):839–843
Romero C, Benedí J, Villar A, Martín-Aragón S (2010) Involvement of Hsp70, a stress protein, in
    the resistance of long-term culture of PC12 cells against sodium nitroprusside (SNP)-induced
    cell death. Arch Toxicol 84:699–708
Rosette C, Karin M (1995) Cytoskeletal control of gene expression: depolymerization of micro-
    tubules activates NF-kappa B. Cytoskeletal control of gene expression: depolymerization of
    microtubules activates NF-kappa B. J Cell Biol 128(6):1111–1119
Rogue PJ, Ritz MF, Malviya AN (1993) Impaired gene transcription and nuclear protein kinase C
    activation in the brain and liver of aged rats. FEBS Lett 334(3):351–354
Ruchalski K, Mao H, Li Z et al (2006) Distinct hsp70 domains mediate apoptosis-inducing factor
    release and nuclear accumulation. J Biol Chem 281:7873–7880
Saleh A, Srinivasula SM, Balkir L, Robbins PD, Alnemri ES (2000) Negative regulation of the
    Apaf-1 apoptosome by Hsp70. Nat Cell Biol 2(8):476–483
Sakahira H, Nagata S (2002) Co-translational folding of caspase-activated DNase with Hsp70,
    Hsp40, and inhibitor of caspase-activated DNase. J Biol Chem 277(5):3364–3370
Schaupp A, Marcinowski M, Grimminger V, Bösl B, Walter S (2007) Processing of proteins by
    the molecular chaperone Hsp104. J Mol Biol 370(4):674–686
Schlieker C, Weibezahn J, Patzelt H et al (2004) Substrate recognition by the AAA + chaperone
    ClpB. Nat Struct Mol Biol 11(7):607–615
Sharma M, Ganguly NK, Chaturvedi G et al (2003) A possible role of HSP70 in mediating car-
    dioprotection in patients undergoing CABG. Mol Cell Biochem 247(1–2):6–31
Shinder GA, Lacourse MC, Minotti S, Durham HD (2001) Mutant Cu/Zn-superoxide dismutase
    proteins have altered solubility and interact with heat shock/stress proteins in models of
    amyotrophic lateral sclerosis. J Biol Chem 276(16):12791–12796
Slee EA, Harte MT, Kluck RM et al (1999) Ordering the cytochrome c-initiated caspase cascade:
    hierarchical activation of caspases-2, -3, -6, -7, -8, and -10 in a caspase-9-dependent manner.
    J Cell Biol 144(2):281–292
Steel R, Doherty JP, Buzzard K et al (2004) Hsp72 inhibits apoptosis upstream of the mitochon-
    dria and not through interactions with Apaf-1. J Biol Chem 279(49):51490–51499

Stankiewicz AR, Lachapelle G, Foo CP, Radicioni SM, Mosser DD (2005) Hsp70 inhibits heat-induced apoptosis upstream of mitochondria by preventing Bax translocation. J Biol Chem 280(46):38729–38739

Suzuki K, Murtuza B, Sammut IA et al (2002) Heat shock protein 72 enhances manganese super-oxide dismutase activity during myocardial ischemia-reperfusion injury, associated with mitochondrial protection and apoptosis reduction. Circulation 106(12 Suppl 1):I270–I276

Sun Y, Ouyang YB, Xu L et al (2006) The carboxyl-terminal domain of inducible Hsp70 protects from ischemic injury in vivo and in vitro. J Cereb Blood Flow Metab 26(7):937–950

Szenczi O, Kemecsei P, Miklós Z et al (2005) In vivo heat shock preconditioning mitigates calcium overload during ischaemia/reperfusion in the isolated, perfused rat heart. Pflugers Arch 449(6):518–525

Tsuchiya D, Hong S, Matsumori Y et al (2003) Overexpression of rat heat shock protein 70 is associated with reduction of early mitochondrial cytochrome C release and subsequent DNA fragmentation after permanent focal ischemia. J Cereb Blood Flow Metab 23(6):718–727

Thorburn A (2004) Death receptor-induced cell killing. Cell Signal 16(2):139–144

Urushitani M, Kurisu J, Tateno M et al (2004) CHIP promotes proteasomal degradation of familial ALS-linked mutant SOD1 by ubiquitinating Hsp/Hsc70. J Neurochem 90(1):231–244

Weibezahn J, Tessarz P, Schlieker C et al (2004) Thermotolerance requires refolding of aggregated proteins by substrate translocation through the central pore of ClpB. Cell 119(5):653–655

Xu L, Giffard RG (1997) HSP70 protects murine astrocytes from glucose deprivation injury. Neurosci Lett 224(1):9–12

Yuan J, Yankner BA (2000) Apoptosis in the nervous system. Nature 407(6805):802–809

Zietkiewicz S, Lewandowska A, Stocki P, Liberek K (2006) Hsp70 chaperone machine remodels protein aggregates at the initial step of Hsp70-Hsp100-dependent disaggregation. J Biol Chem 281(11):7022–7099

Chapter 4
Mechanisms of Activation and Inactivation of HSP70 Synthesis

Abstract The main participants that regulate HSP70 synthesis are the HSP70 genes and the HSF-1 transcription factor. In a non-stressed cell, HSF-1 exists within the cytoplasm in an inactive monomer state. The inactive state of the HSF-1 monomer is supported by the «HSP90-p23-immunophilin» complex and, possibly, by intramolecular hydrophobic bonds and phosphorylation of specific HSF-1 serine residues. HSP70 and HSP40 participate in the formation of the inhibitory complex. Activation of the HSF-1 transcription factor occurs in two steps. The first step occurs when emerging denatured and/or misfolded proteins induce removal of the negative influence of the «HSP90-p23-immunophylin» complex, trimerization of the HSF-1, and the binding of trimerized HSF-1 to a specific region of the *hsp70* promoter, termed the HSE. In the second step, the negative influence of the «HSP90-p23-Fkbp52» complex is abolished and transcription is activated. Inactivation of HSF-1 and cessation of HSP70 synthesis occurs when HSP70 and HSP40 bind to HSF-1, and thus inhibit HSF-1 transcriptional activity. Thus, the system of HSP70 synthesis includes an autoregulation mechanism; the ability of HSP70 to inactivate its own transcription factor.

Keywords HSP70 • HSE • HSF-1 • The HSP70 synthesis

In the last chapter we examined HSP70 functions in a damaged cell. We learnt that: (i) to maintain protein homeostasis the FORD protein quality control system prevents protein aggregation, disaggregates the formed aggregates and degrades misfolded proteins that cannot be retrieved to an active state; (ii) as a result of HIF-1 activation, restoration of protein homeostasis forms a specific cell defence mechanism against hypoxic, free-radical and calcium injuries; (iii) HSP70 can deposit mutant proteins and (iv) HSP70 inhibits apoptosis and protects against the accidental triggering of apoptosis in healthy cell.

These points allowed us to make an important conclusion: HSP70 is a component of an intracellular system aimed at maintaining protein homeostasis and the recovery of damaged cells. Evidence for the protective role of HSP70 was obtained in many studies (Huot et al. 1991; Jaattela et al. 1992; Parsell and

I. Malyshev, *Immunity, Tumors and Aging: The Role of HSP70*,
SpringerBriefs in Biochemistry and Molecular Biology,
DOI: 10.1007/978-94-007-5943-5_4, © The Author(s) 2013

Lindquist 1993; Marber et al. 1995; Mestril et al. 1994; Mehlen et al. 1995; Mizzen and Welch 1988; Mosser et al. 1997; Plumier et al. 1995).

At the same time, with an understanding of the key roles HSP70 plays in the process of cell reparation and cell defence, another important question emerged: what are the mechanisms regulating HSP70 synthesis? This chapter will address this question.

4.1 The Structure of Genes and HSP70 Transcription Factors

The main participants in the regulatory mechanisms for HSP70 synthesis are *hsp70* genes, forming a whole family; HSF transcription factors are another family (HSF-1 will be of particular interest for us); and finally the regulatory mechanisms of HSP70 synthesis, which remain unclear.

The human genome contains 17 genes that encode different isoforms of HSP70. HSP70 genes are localized in 11 chromosomes 1, 4, 5, 6, 9, 10, 11, 13, 14, 20 and 21 (Brocchieri et al. 2008).

Figure 4.1 schematically shows the gene encoding HSP70. As with all genes, the *hsp70* gene contains two main parts: the promoter and the encoding sequence. The promoter contains the TATA box and a marker regulatory nucleotide sequence for the *hsp70* gene, the heat shock consensus element (HSE). RNA polymerase recognizes and binds to the TATA box. This part of the promoter is nonspecific and is present in all known genes. HSE is a specific unit of the *hsp70* gene promoter. HSE contains many adjacent and inverted repeats of the pentanucleotide 5′-nGAAn-3′ (Fernandes et al. 1994). HSE is necessary for activation of the *hsp70* gene, because factors activating HSF-1 transcription bind precisely to HSE. HSF-1 binding to HSE causes activation of RNA polymerase and, consequently, gene transcription (Lis and Wu 1993; Morimoto 1993; Voellmy 1994; Wu 1995).

By the mid-1990s it was already well known what HSFs existed and their structures had been solved. Multicellular invertebrates have only one kind of HSF, whereas vertebrates have four different types (Clos et al. 1990; Czarnecka-Verner et al. 1995; Nakai and Morimoto 1993; Nakai et al. 1997; Nover et al. 1996; Scharf et al. 1990; Scharf et al. 1993; Schuetz et al. 1991; Sorger and Pelham 1987; Treuter et al. 1993; Wiederrecht et al. 1988). HSF-1, HSF2 and HSF4 were discovered in cells of all mammals, whereas HSF3 exists only in birds.

Fig. 4.1 The gene encoding HSP70

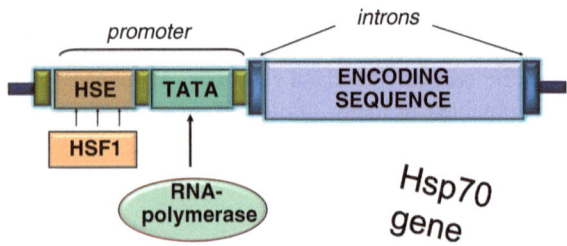

HSF-1 is a key factor in the stress-induced gene expression in most vertebrates, except birds (Ali et al. 1998; Baler et al. 1993; Holmberg et al. 2001; McMillan et al. 1998; Sarge et al 1993; Zhang et al. 2002).

Figure 4.2 shows the domain structure and functional aspects of HSF-1. HSF-1 has a DNA-binding domain (Harrison et al. 1994; Schultheiss et al. 1996; Vuister et al. 1994) that is located near the amino terminus (Clos et al. 1990). This domain interacts with HSE in the *hsp70* gene promoter. The DNA-binding domain is adjacent to a long domain with the hydrophobic repeat HR-A/B. This domain is necessary for HSF trimerization (Clos et al. 1990; Peteranderl and Nelson 1992; Sorger and Nelson 1989) and trimerization is essential for facilitating DNA-binding. The C-terminal region contains a transactivational domain that is required for the activation of transcription. (Chen et al. 1993; Green et al. 1995; Shi et al. 1995; Zuo et al. 1995; Wisniewski et al. 1996). Near the transactivational domain there is another sequence of hydrophobic repeats, HR-C. In-between HR-A/B and HR-C there is a sequence that, together with these sites, suppresses DNA-binding and the transcriptional activity of HSF-1 (Farkas et al. 1998; Green et al. 1995; Hoj and Jakobsen 1994; Nieto-Sotelo et al. 1990; Orosz et al. 1996; Rabindran et al. 1993a, b; Shi et al. 1995; Zuo et al. 1994; 1995).

4.2 The Mechanism for Activation of HSP70, or the Story of How Two Richards Argued

There was no doubt about the general course of events in the activation of HSP70 synthesis: it was clear that HSF-1 in normal cells exists in the inactive form and cannot bind DNA to activate gene transcription. Under certain conditions, HSF-1 can be activated. Activation of HSF-1 leads to interaction of this

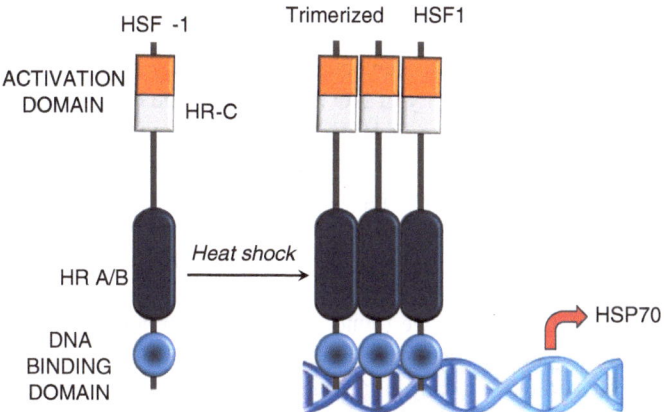

Fig. 4.2 The domain structure and functional aspects of HSF-1 (heat shock transcription factor 1)

protein with the *hsp70* gene promoter and subsequent transcription of the *hsp70* gene. The inactive form of HSF-1 is a monomer. When HSF-1 is activated, the transcription factor trimerizes to form a single active unit. In the trimer state, HSF-1 acquires the ability to penetrate into the nucleus and bind to the *hsp70* gene promoter.

One remarkable property of HSP70 is that in the first few hours following stress the overall biosynthesis of cellular proteins is significantly depressed, whereas the synthesis of HSP70 dramatically increases (Fig. 1.2) and the accumulation of HSP70 proteins can reach a maximum after 2–3 days. Conversely, when general protein biosynthesis recovers, HSP70 synthesis decreases and is eventually terminated! This indicates that the regulation of HSP70 synthesis differs from the regulation of synthesis of most cell proteins.

That was a true challenge to molecular biologists. Many ambitious researchers started dealing with the problem and contributed a lot to solving it. First of all, it was important to find out what triggers the activation of HSF-1, i.e., whether HSF-1 directly perceives physical and chemical changes in a stressed cell or whether there are any intermediaries between the stress signal and HSF-1.

The question of whether HSF-1 activation can be induced directly by physical conditions or chemical factors has been periodically emerging and has left researchers baffled. Goodson and Sarge 1995, Larson et al. 1995 and Zhong et al. 1998 were among those who spent considerable time trying to understand this problem. They showed that in vitro purified or recombinant HSF-1 in the monomeric form can be activated independently, i.e., they form trimeric structures and acquire DNA-binding activity in response to heat or hydrogen peroxide stress (Goodson and Sarge 1995; Larson et al. 1995; Zhong et al. 1998), in response to lower pH (pH 6.5) or to the addition of salicylates (Mosser et al. 1990; Zhong et al. 1998) in vitro. These results suggest that intrinsic properties of HSF-1 can be modulated by its chemical environment.

Some time ago I too was quite interested in the issue. Together with Prof. Manukhina and Prof. Vanin, we ran experiments on cultured cells and showed that a nitric oxide (NO) donor caused a marked activation of HSP70 synthesis (Malyshev et al. 1996a, b; Wiegant et al. 1999). We suggested that this was because of the well-known ability of NO to form disulfide bonds (Fig. 4.3). It cannot be excluded that in the case of HSF-1, NO nitrosylated SH groups of HSF-1 and thereby induces HSF-1 trimerization and activation. Ahn and Thiele (2003) confirmed our hypothesis by showing that oxidation of two cysteines could result in the formation of disulfide bonds both in vitro and in vivo under the action of heat shock or HSF-1 oxidants. These bonds may possibly contribute to the trimerization and activation of HSF-1.

However, as often happens, we found a fly in the ointment. A few experiments showed that the cysteines that may have formed disulfide bonds between the HSF-1 molecules do not play any significant role in the trimerization since mutant HSF-1 without these cysteines still had the capability for trimerization under heat shock (Orosz et al. 1996; Rabindran et al. 1993a, b; Zuo et al. 1994, 1995).

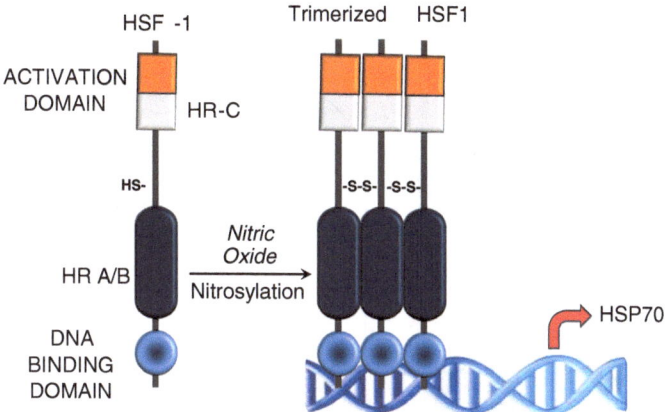

Fig. 4.3 Our hypothesis: NO nitrosylated SH groups of HSF-1 and thereby induces HSF-1 trimerization and activation

Therefore, many researchers arrived at an idea that was common for a variety of HSP70 synthesis inducers in vivo; that all inducers caused the unfolding and accumulation of unfolded or damaged proteins (Freeman et al. 1995; Liu et al. 1996; McDuffee et al. 1997; Senisterra et al. 1997; Zou et al. 1998). Consequently, an increase in the level of denatured unfolded proteins can possibly be a trigger for HSF-1 activation (Kelley and Schlesinger 1978; Hightower 1980).

As soon as this idea emerged, it was quickly put to the test! How? It was very simple! Denatured proteins were administered to cells and it worked! The cells responded immediately by activation of HSF-1 and subsequent synthesis of HSP70 (Ananthan et al. 1986). These observations clearly indicated that denatured proteins triggered activation of HSF-1 and, consequently, HSP synthesis.

Next, it was important to understand how cells translate the information about the appearance of denatured proteins to HSF-1 activation mechanisms. The answer was right in front of us. Researchers who have studied HSP70 induction certainly knew very well that it was the heat shock protein that could best identify and bind denatured proteins. Further, researchers reasoned as follows: if HSP binding to denatured proteins in a damaged cell leads to HSF activation, then one could assume that HSPs were bound to HSF-1 in normal cells, and HSF activation was a result of HSP detachment from HSF-1.

Furthermore, Westwood and colleagues discovered that a heat shock to *Drosophila* cells induced HSF homotrimerization (Westwood et al. 1991; Westwood and Wu 1993). Subsequent studies on other multicellular animals have confirmed this phenomenon (Baler et al. 1993; Sarge et al. 1993). They also showed that homotrimerization induced HSF-1 binding to DNA/HSE (Zuo et al. 1994). It immediately became clear that the trimerized DNA-bound HSF-1 can exist in both transcriptionally active and transcriptionally inactive states (Bruce et al. 1993; Jurivich et al. 1992). Therefore, it also became clear that HSF-1

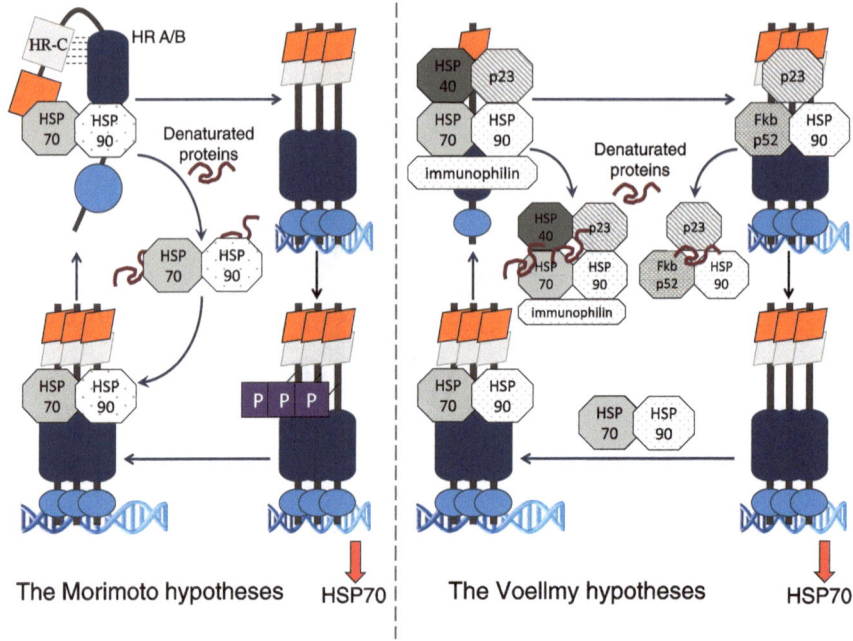

Fig. 4.4 Richard Morimoto's hypothesis versus Richard Voellmy's hypothesis: the two researchers disagreed on the manner in which the HSF-1 activation were regulated

activation had at least two phases and that the phase of acquiring the DNA-binding activity and the phase of acquiring the transcriptional activity were regulated by separate mechanisms (Zuo et al. 1995). This finding was confirmed by the HSF-1 domain structure (Fig. 4.2): it is the HS-A/B domain that is closer to the N-terminus and is responsible for the trimerization, whereas the domain that is responsible for the transactivational activity is closer to the C-terminus. (Knauf et al. 1996; Zuo et al. 1995).

However, we have approached a real understanding of this issue thanks to two Richards: Richard Morimoto and Richard Voellmy. When you read their papers (Morimoto 1998; Voellmy 1994; Ananthan et al. 1986), you have a real impression of a true scientific competition; the discussion polite in form, but fierce in essence. They were rivals, but unwittingly helped each other. As with two athletes, when both are running together each one runs faster than if he was running alone. The two researchers disagreed on the manner in which the first and the second steps of HSF-1 activation were regulated (Fig. 4.4).

Morimoto believed that in normal cells HSF-1 exists in an inactive monomeric state due to intramolecular interactions between HR-C, HR-A/B and the HSF-1 central region (Nakai and Morimoto 1993; Rabindran et al. 1993a, b), as well as due to phosphorylation of serine residues Ser_{303}, Ser_{307}, and Ser_{363} (Knauf et al. 1996; Kline and Morimoto 1997; Holmberg et al. 2002). Stabilization of intramolecular hydrophobic interactions, in this case, was ensured by HSP70 and HSP90. When denatured proteins appeared in the cell, HSP70 and HSP90 detached themselves from HSF-1 and formed complexes with the emerging damaged proteins. This led to destabilization of intra-molecular hydrophobic interactions that prevented HSF-1 trimerization. As a result, HSF-1 spontaneously trimerized and acquired the ability to bind HSE. In the next stage, according to Morimoto, the stress-induced hyperphosphorylation of additional residues Ser_{230} and Thr_{142} took place and this led to the appearance of transcriptional activity (Cotto et al. 1996; Hensold et al. 1990; Holmberg et al. 2001; Jurivich et al. 1992; Kline and Morimoto 1997; Sarge et al. 1993; Soncin et al. 2003). Indeed it was discovered that protein kinases Erk1/2, Gsk3, Jnk, CamkII, Rsk2, Ck2 and Pkc phosphorylated HSF-1. However, no direct evidence for intramolecular hydrophobic interactions has been observed.

Voellmy proposed an alternative explanation (Fig. 4.4). In essence his hypothesis suggested that the HSP90-chaperone complex plays a key role in the regulation of trimerization and the appearance of HSF-1 transcriptional properties.

The following experimental facts provided the proof for Voellmy's explanation. First of all, Hsp90 and co-chaperones can interact with HSF-1 (Nadeau et al. 1993; Nair et al. 1996). Secondly, Hsp90-binding medicines herbimycin A and geldanamycin activate HSF-1 (Hedge et al. 1995; Zou et al. 1998). Thirdly, removing HSP90 but not other chaperones, (Zou et al. 1998) or adding HSP90 antibodies (Ali et al. 1998) induced the DNA-binding activity in HSF-1, whereas the administration of purified HSP90 abolished this effect. Finally, excessive HSP90 synthesis reduced the heat-shock-induced induction of HSF-1 DNA-binding activity (Zhao et al. 2002). Interpretation of these results converged to one conclusion: HSP90 inhibits HSF-1 activation.

Evidence for the role of a p23 protein (Bharadwaj et al. 1999) and immunophilin (Duina et al. 1998) in the inhibition of the DNA-binding properties of HSF-1 was obtained in the same way.

Thus, oligomerization of HSF-1 and its DNA-binding activity can, apparently, be inhibited through association of monomeric HSF-1 with the HSP90 multi-chaperone complex. If so, then such HSP90-HSF-1 complexes should be found in non-stressed cells and should disappear under stress at a rate comparable to the rate of trimerization. Research in the Voellmy laboratory proved that this was exactly what happened in cells.

By then other researchers, using *Drosophila* cells, showed that removal of not only HSP90, but also of HSP70 and HSP40 increased the DNA-binding activity of HSF-1 (Marchler and Wu 2001). Together all these data allowed Voellmy to propose his model of HSF-1 activation (Fig. 4.4).

Initially, during the synthesis in ribosomes, monomeric HSF-1 binds HSP70 and HSP40. Then these proteins form the multi-chaperone «HSP90-p23-immunnofilin» complex, which reliably stabilizes the monomeric inactive state of HSF-1 in a non-stressed cell. When a cell experiences a stress response, denatured proteins appear that "distract" the chaperones and co-chaperones that are part of the HSF-1 inhibitory complex. As a result, the HSF-1 molecules are released from the inhibitory effect of the chaperone complex and form homotrimers with DNA-binding activity.

Then the HSF-1 trimer binds to DNA, and simultaneously the regulatory domain of HSF-1 binds to another HSP90-multi-chaperone complex «Hsp90-p23-Fkbp52» (Guo et al. 2001; Nair et al. 1996), which inhibits HSF-1 transcriptional activity. If the cell has significantly high levels of denatured proteins, then these misfolded proteins will also "distract" this inhibitory complex, thus removing the second constraint and allow HSF-1 to activate *hsp70* gene transcription and synthesis.

However, data that showed that not all HSF-1 molecules are associated with the multi-chaperone complex in a non-stressed cell introduced doubt into the elegant Voellmy model. Therefore we cannot exclude that the inactive condition of a part of the HSF-1 population may be supported by other mechanisms, different from those based on chaperones, for instance by the hydrophobic mechanism suggested by Morimoto. In this way, the Morimoto and Voellmy hypotheses no longer "competed", each one simply showing alternative ways of supporting the inactive state of HSF-1.

4.3 The Mechanism of Inactivating the HSP70 Synthesis, or the Story About How Two Richards Agreed

HSP70 synthesis stops when stress recedes and protein homeostasis in the cell recovers. Researchers did not disagree about the inactivation of HSF-1 and the subsequent decrease of HSP70 synthesis. Here, agreement was reached that upon recession of cell stress, HSP70 and HSP40 bind to the trimerized HSF-1, or more specifically, with its transactivational domain (Abravaya et al. 1992; Baler et al. 1992; Halladay and Craig 1995; Mosser et al. 1993; Shi et al. 1995; Shi et al. 1998), Together with another protein, HSBP1 (HSF binding protein 1) (Satyal et al. 1998) they induce dissociation of HSF-1 from DNA to return HSF-1 to the monomeric inactive form. Both Richard Morimoto and Richard Voellmy agreed with such explanations.

Another researcher, Boellmann, introduced another intriguing point on the mechanism of HSP70 synthesis. He proved that the Daxx protein is an important co-factor in activating the transcriptional activity of HSF-1 (Boellmann et al. 2004). Until now, it remains unclear how Daxx increases the transcriptional activity of HSF-1. It could be competing with the Hsp90-p23-Fkbp52 chaperone complex for binding to the trimer HSF-1 sites. At the same time there is another interesting point! Initially Daxx was described as an amplifier of apoptosis and a repressor of the general transcriptional activity (Michaelson 2000). It is amazing how the activation of the same intracellular protein can lead both to apoptosis, i.e., the death program, or to activation of HSP70 synthesis—the life program! Figure 4.5 clearly illustrates this paradox—the lack of boundaries between life and death.

Fig. 4.5 "Life and death"

This picture is from the Mark Twain museum in Virginia. Depending on how you look at the picture, with eyes wide open or squinting, you will see two completely opposite things: a skull—a symbol of death, or children—a symbol of life! In science, as in life, a lot depends on your view of the world.

4.4   Regulation of HSP70 Synthesis, the Protein Which Supports Protein Homeostasis in a Cell, is Itself Based on the Homeostatic Principle

We can make another important conclusion now. It is certainly very interesting that HSP70 is involved in the mechanism for controlling its expression. However, it hides something greater than the mechanisms of substrate inhibition, which are well-known in biochemistry! The activation of HSP70 synthesis occurs when denatured proteins appear, i.e., when homeostasis inside the cell is disrupted. The existing fraction of HSP70, which inhibits HSF-1, detects the damaged and denatured proteins and binds to them. This HSP70, dissociating from HSF-1, enables activation of the transcription factor HSF-1. The active HSF-1 enters the nucleus and activates expression of *hsp70* genes, which, in turn, increases the synthesis of HSP70. The newly synthesized HSP70, using the ATP-ase cycle, either sends damaged proteins for refolding, or targets irreversibly damaged proteins to proteasomes or lysosomes, i.e., HSP70 restores the disturbed homeostasis. Only after reaching homeostasis once more in the cell, does the synthesis of HSP70 stop. So this is a ready-made mechanism of homeostatic negative feedback! Indeed, the pre-existing HSP70 fraction serves as a sensor, detecting a disorder in protein home-ostasis and sending a signal to the transcription factor HSF-1. HSF-1 and *hsp70* genes function as control switches, regulating cellular responses to the disruption of protein homeostasis. Finally, the newly synthesized HSP70 functions as effec-tors of a homeostatic negative feedback mechanism which restores homeostasis. A possibility that such a negative feedback mechanism of could exist was sug-gested many years ago by Lindquist (1980) and Didomenico et al. (1982).

Now we can expand our notion of the homeostatic control mechanism for pro-tein turnover, which we discussed during the third chapters (Fig. 3.1), by intro-ducing new elements including HSF-1, the HSP90-chaperone complex, the *hsp70* gene, possible mechanisms of HSF-1 phosphorylation, and some other additional regulators, such as Daxx.

4.5 What New Have we Learnt About the Mechanisms of HSP70 Synthesis (Summary) and P.S

We can summarize what we have learnt about the mechanisms of HSP70 synthesis.

1. The regulatory element, the HSE promoter, plays a key role in the transcrip-tion of the *hsp70* gene. The interaction of the transcription factor HSF-1 with HSE activates the expression of *hsp70* gene.
2. In a non-stressed cell, HSF-1 is present in the cytoplasm in an inactive monomeric state. The inactive state of the HSF-1 monomer is supported by the « HSP90-p23-immunophilin » complex and, possibly, by intramolecular hydrophobic bonds and phosphorylation of specific HSF-1 serine residues. HSP70 and HSP40 participate in the formation of the inhibitory complex.
3. Activation of the HSF-1 transcription factor occurs via two steps. In the first step, emerging denatured proteins induce removal of the negative influence of the «HSP90-p23-immunophilin» complex, which leads to trimerization HSF-1, and this trimer binds DNA. In the second step, the negative influence of the «HSP90-p23-Fkbp52» complex is abolished and transcription is activated.
4. Inactivation of HSF-1 and cessation of HSP70 synthesis are facilitated by the interaction of HSP70 and HSP40 with HSF-1. Such an interaction inhibits the transcriptional activity of HSF-1 by facilitating dissociation of HSF-1 with DNA.
5. The HSP70 synthesis system includes an autoregulation mechanism; based on the ability of HSP70 to inactivate its own transcription factor.

P.S. When I was trying to figure out how the mechanisms of HSP70 synthesis work, I do not know why, but I kept thinking about stories from the Bible. As you know, according to the New Testament, Christ suffered on the cross for 3 days. During these days he was actually experiencing the most severe psycho-emotional stress, pain, as well as dehydration and heat shock. With this in mind, I thought that over this period his body must have accumulated a significant level of HSP70. So, when I learnt that the maximum stress-induced accumulation of HSP70 can happen on the third day, i.e., the day when Christ was miraculously resurrected, I started thinking, was it just by chance? Moreover, when I remembered about the antiapoptic ability of HSP70, i.e., the ability to defeat death in a cell, I suddenly thought: could these proteins be involved in the resurrection of Christ!?

However, do not rush to accuse me of naively trying to unravel the nature of naive biblical miracles! I remember well that the two others, hanging on the cross next to Jesus Christ, did not resurrect! The mystery remained a mystery! However, remember, in Chap. 2 I introduced to you how the HSP70-dependent system of re-folding and degradation sorts out damaged proteins into those that are liable for recovery—they are sent to refolding-"resurrection"—while the irreversibly damaged proteins are sent for degradation. But those two, who were next to Christ on crosses, were incorrigible criminals.

References

Abravaya K, Myers MP, Murphy SP, Morimoto RI (1992) The human heat shock protein hsp70 interacts with HSF, the transcription factor that regulates heat shock gene expression. Genes Dev 6(7):1153–1164

Ahn SG, Thiele DJ (2003) Redox regulation of mammalian heat shock factor 1 is essential for Hsp gene activation and protection from stress. Genes Dev 17(4):516–528

Ali A, Bharadwaj S, O'Carroll R, Ovsenek N (1998) HSP90 interacts with and regulates the activity of heat shock factor 1 in Xenopus oocytes. Mol Cell Biol 18(9):4949–4960

Ananthan J, Goldberg AL, Voellmy R (1986) Abnormal proteins serve as eukaryotic stress signals and trigger the activation of heat shock genes. Science 4749:522–524

Baler R, Welch WJ, Voellmy R (1992) Heat shock gene regulation by nascent polypeptides and denatured proteins: hsp70 as a potential autoregulatory factor. J Cell Biol 117(6):1151–1159

Baler R, Dahl G, Voellmy R (1993) Activation of human heat shock genes is accompanied by oligomerization, modification, and rapid translocation of heat shock transcription factor HSF1. Mol Cell Biol 13(4):2486–2496

Boellmann F, Guettouche T, Guo Y, Fenna M, Mnayer L, Voellmy R (2004) DAXX interacts with heat shock factor 1 during stress activation and enhances its transcriptional activity. Proc Natl Acad Sci USA 101(12):4100–4105

Bharadwaj S, Ali A, Ovsenek N (1999) Multiple components of the HSP90 chaperone complex function in regulation of heat shock factor 1 In vivo. Mol Cell Biol 19(12):8033–8041

Brocchieri L, Conway de Macario E, Macario AJ (2008) Hsp70 genes in the human genome: Conservation and differentiation patterns predict a wide array of overlapping and specialized functions. BMC Evol Biol 8:19

Bruce JL, Price BD, Coleman CN, Calderwood SK (1993) Oxidative injury rapidly activates the heat shock transcription factor but fails to increase levels of heat shock proteins. Cancer Res 53(1):12–15

Chen Y, Barlev NA, Westergaard O, Jakobsen BK (1993) Identification of the C-terminal activator domain in yeast heat shock factor: independent control of transient and sustained transcriptional activity. EMBO J 12(13):5007–5018

Clos J, Westwood JT, Becker PB, Wilson S, Lambert K, Wu C (1990) Molecular cloning and expression of a hexameric Drosophila heat shock factor subject to negative regulation. Cell 63(5):1085–1097

Cotto JJ, Kline M, Morimoto RI (1996) Activation of heat shock factor 1 DNA binding precedes stress-induced serine phosphorylation. Evidence for a multistep pathway of regulation. J Biol Chem 271(7):3355–3358

Czarnecka-Verner E, Yuan CX, Fox PC, Gurley WB (1995) Isolation and characterization of six heat shock transcription factor cDNA clones from soybean. Plant Mol Biol 29(1):37–51

DiDomenico BJ, Bugaisky GE, Lindquist S (1982) The heat shock response is self-regulated at both the transcriptional and posttranscriptional levels. Cell 31(3 Pt 2):593–603

Duina AA, Kalton HM, Gaber RF (1998) Requirement for Hsp90 and a CyP-40-type cyclophilin in negative regulation of the heat shock response. J Biol Chem 273(30):18974–18978

Farkas T, Kutskova YA, Zimarino V (1998) Intramolecular repression of mouse heat shock factor 1. Mol Cell Biol 18(2):906–918

Freeman ML, Borrelli MJ, Syed K, Senisterra G, Stafford DM, Lepock JR ye фд (1995) Characterization of a signal generated by oxidation of protein thiols that activates the heat shock transcription factor. J Cell Physiol 164(2):356–366

Fernandes M, Xiao H, Lis JT (1994) Fine structure analyses of the Drosophila and Saccharomyces heat shock factor-heat shock element interactions. Nucleic Acids Res 22(2):167–173

Green M, Schuetz TJ, Sullivan EK, Kingston RE (1995) A heat shock-responsive domain of human HSF1 that regulates transcription activation domain function. Mol Cell Biol 15(6):3354–3362

Goodson ML, Sarge KD (1995) Heat-inducible DNA binding of purified heat shock transcription factor 1. J Biol Chem 270(6):2447–2450

Guo Y, Guettouche T, Fenna M et al (2001) Evidence for a mechanism of repression of heat shock factor 1 transcriptional activity by a multichaperone complex. J Biol Chem 276(49):45791–45799

Halladay JT, Craig EA (1995) A heat shock transcription factor with reduced activity suppresses a yeast HSP70 mutant. Mol Cell Biol 15(9):4890–4897

Harrison CJ, Bohm AA, Nelson HC (1994) Crystal structure of the DNA binding domain of the heat shock transcription factor. Science 263(5144):224–227

Hegde RS, Zuo J, Voellmy R, Welch WJ (1995) Short circuiting stress protein expression via a tyrosine kinase inhibitor, Herbimycin A. J Cell Physiol 165(1):186–200

Hensold JO, Hunt CR, Calderwood SK, Housman DE, Kingston RE (1990) DNA binding of heat shock factor to the heat shock element is insufficient for transcriptional activation in murine erythroleukemia cells. Mol Cell Biol 10(4):1600–1608

Hightower LE (1980) Cultured animal cells exposed to amino acid analogues or puromycin rapidly synthesize several polypeptides. J Cell Physiol 3:407–427

Høj A, Jakobsen BK (1994) A short element required for turning off heat shock transcription factor: evidence that phosphorylation enhances deactivation. EMBO J 13(11):2617–2624

Holmberg CI, Hietakangas V, Mikhailov A et al (2001) Phosphorylation of serine 230 promotes inducible transcriptional activity of heat shock factor 1. EMBO J 20(14):3800–3810

Holmberg CI, Tran SE, Eriksson JE, Sistonen L (2002) Multisite phosphorylation provides sophisticated regulation of transcription factors. Trends Biochem Sci 27(12):619–627

Huot C, Tremblay J, Hamet P (1991) Cell biology of atrial natriuretic peptide. Blood Vessels 28(1–3):84–92

Jäättelä M, Wissing D, Bauer PA, Li GC (1992) Major heat shock protein hsp70 protects tumor cells from tumor necrosis factor cytotoxicity. EMBO J 11(10):3507–3512

Jurivich DA, Sistonen L, Kroes RA, Morimoto RI (1992) Effect of sodium salicylate on the human heat shock response. Science 255(5049):1243–1245

Kelley PM, Schlesinger MJ (1978) The effect of amino acid analogues and heat shock on gene expression in chicken embryo fibroblasts. Cell 15(4):1277–1286

Kline MP, Morimoto RI (1997) Repression of the heat shock factor 1 transcriptional activation domain is modulated by constitutive phosphorylation. Mol Cell Biol 4:2107–2115

Knauf U, Newton EM, Kyriakis J, Kingston RE (1996) Repression of human heat shock factor 1 activity at control temperature by phosphorylation. Genes Dev 10(21):2782–2793

Larson JS, Schuetz TJ, Kingston RE (1995) In vitro activation of purified human heat shock factor by heat. Biochem 34(6):1902–1911

Lindquist S (1980) Varying patterns of protein synthesis in Drosophila during heat shock: implications for regulation. Dev Biol 77(2):463–479

Lis J, Wu C (1993) Protein traffic on the heat shock promoter: parking, stalling, and trucking along. Cell 74(1):1–4

Liu H, Lightfoot R, Stevens JL (1996) Activation of heat shock factor by alkylating agents is triggered by glutathione depletion and oxidation of protein thiols. J Biol Chem 271(9):4805–4812

Malyshev IYu, Malugin AV, Golubeva LYu et al (1996a) Nitric oxide donor induces HSP70 accumulation in the heart and in cultured cells. FEBS Lett 391(1–2):21–23

Malyshev IYu, Malugin AV, Manukhina EB et al (1996b) Is HSP70 involved in nitric oxide-induced protection of the heart? Physiol Res 45(4):267–272

Marchler G, Wu C (2001) Modulation of Drosophila heat shock transcription factor activity by the molecular chaperone DROJ1. EMBO J 20(3):499–509

Marber MS, Mestril R, Chi SH, Sayen MR, Yellon DM, Dillmann WH (1995) Overexpression of the rat inducible 70-kD heat stress protein in a transgenic mouse increases the resistance of the heart to ischemic injury. J Clin Invest 95(4):1446–1456

McDuffee AT, Senisterra G, Huntley S et al (1997) Proteins containing non-native disulfide bonds generated by oxidative stress can act as signals for the induction of the heat shock response. J Cell Physiol 171(2):143–151

McMillan DR, Xiao X, Shao L, Graves K, Benjamin IJ (1998) Targeted disruption of heat shock transcription factor 1 abolishes thermotolerance and protection against heat-inducible apoptosis. J Biol Chem 273(13):7523–7528

Mehlen P, Kretz-Remy C, Briolay J et al (1995) Intracellular reactive oxygen species as apparent modulators of heat-shock protein 27 (hsp27) structural organization and phosphorylation in basal and tumour necrosis factor alpha-treated T47D human carcinoma cells. Biochem J 312 (Pt 2):367–375

Mestril R, Chi SH, Sayen MR, Dillmann WH (1994) Isolation of a novel inducible rat heat-shock protein (HSP70) gene and its expression during ischaemia/hypoxia and heat shock. Biochem J 298(Pt 3):561–569

Michaelson JS (2000) The Daxx enigma. Apoptosis 5(3):217–220

Mizzen LA, Welch WJ (1988) Characterization of the thermotolerant cell. I. Effects on protein synthesis activity and the regulation of heat-shock protein 70 expression. J Cell Biol 106(4):1105–1116

Morimoto RI (1993) Cells in stress: transcriptional activation of heat shock genes. Science 259(5100):1409–1410

Morimoto RI (1998) Regulation of the heat shock transcriptional response: cross talk between a family of heat shock factors, molecular chaperones, and negative regulators. Genes Dev 12:3788–3796

Mosser DD, Kotzbauer PT, Sarge KD, Morimoto RI (1990) In vitro activation of heat shock transcription factor DNA-binding by calcium and biochemical conditions that affect protein conformation. Proc Natl Acad Sci USA 87(10):3748–3752

Mosser DD, Duchaine J, Massie B (1993) The DNA-binding activity of the human heat shock transcription factor is regulated in vivo by hsp70. Mol Cell Biol 13(9):5427–5438

Mosser DD, Caron AW, Bourget L, Denis-Larose C, Massie B (1997) Role of the human heat shock protein hsp70 in protection against stress-induced apoptosis. Mol Cell Biol 17(9):5317–5327

Nadeau K, Das A, Walsh CT (1993) Hsp90 chaperonins possess ATPase activity and bind heat shock transcription factors and peptidyl prolyl isomerases. J Biol Chem 268(2):1479–1487

Nakai A, Morimoto RI (1993) Characterization of a novel chicken heat shock transcription factor, heat shock factor 3, suggests a new regulatory pathway. Mol Cell Biol 13(4):1983–1997

Nakai A, Tanabe M, Kawazoe Y, Inazawa J, Morimoto RI, Nagata K (1997) HSF4, a new member of the human heat shock factor family which lacks properties of a transcriptional activator. Mol Cell Biol 17(1):469–481

Nair SC, Toran EJ, Rimerman RA et al (1996) A pathway of multi-chaperone interactions common to diverse regulatory proteins: estrogen receptor, Fes tyrosine kinase, heat shock transcription factor Hsf1, and the aryl hydrocarbon receptor. Cell Stress Chaperones 1(4):237–250

Nieto-Sotelo J, Wiederrecht G, Okuda A, Parker CS (1990) The yeast heat shock transcription factor contains a transcriptional. activation domain whose activity is repressed under non-shock conditions. Cell 62(4):807–817

Nover L, Scharf KD, Gagliardi D et al (1996) The Hsf world: classification and properties of plant heat stress transcription factors. Cell Stress Chaperones 1(4):215–223

Orosz A, Wisniewski J, Wu C (1996) Regulation of Drosophila heat shock factor trimerization: global. sequence requirements and independence of nuclear localization. Mol Cell Biol 16(12):7018–7030

Parsell DA, Lindquist S (1993) The function of heat-shock proteins in stress tolerance: degradation and reactivation of damaged proteins. Annu Rev Genet 27:437–496

Peteranderl R, Nelson HC (1992) Trimerization of the heat shock transcription factor by a triple-stranded alpha-helical coiled-coil. Biochem 31(48):12272–12276

Plumier JC, Ross BM, Currie RW et al (1995) Transgenic mice expressing the human heat shock protein 70 have improved post-ischemic myocardial. recovery. J Clin Invest 95(4):1854–1860

Rabindran SK, Haroun RI, Clos J, Wisniewski J, Wu C (1993a) Regulation of heat shock factor trimer formation: role of a conserved leucine zipper. Science 259(5092):230–234

Rabindran SK, Wisniewski J, Li L, Li GC, Wu C (1993b) Interaction between heat shock factor and hsp70 is insufficient to suppress induction of DNA-binding activity in vivo. Mol Cell Biol 14(10):6552–6560

Sarge KD, Murphy SP, Morimoto RI (1993) Activation of heat shock gene transcription by heat shock factor 1 involves oligomerization, acquisition of DNA-binding activity, and nuclear localization and can occur in the absence of stress. Mol Cell Biol 13(3):1392–1407

Satyal SH, Chen D, Fox SG, Kramer JM, Morimoto RI (1998) Negative regulation of the heat shock transcriptional response by HSBP1. Genes Dev 12(13):1962–1974

Scharf KD, Rose S, Zott W, Schöffl F, Nover L (1990) Three tomato genes code for heat stress transcription factors with a region of remarkable homology to the DNA-binding domain of the yeast HSF. EMBO J 9(13):4495–4501

Scharf KD, Rose S, Thierfelder J, Nover L (1993) Two cDNAs for tomato heat stress transcription factors. Plant Physiol 102(4):1355–1356

Schuetz TJ, Gallo GJ, Sheldon L, Tempst P, Kingston RE (1991) Isolation of a cDNA for HSF2: evidence for two heat shock factor genes in humans. Proc Natl Acad Sci USA 88(16):6911–6915

Schultheiss J, Kunert O, Gase U, Scharf KD, Nover L, Rüterjans H (1996) Solution structure of the DNA-binding domain of the tomato heat-stress transcription factor HSF24. Eur J Biochem 236(3):911–921

Senisterra GA, Huntley SA, Escaravage M et al (1997) Destabilization of the Ca2+-ATPase of sarcoplasmic reticulum by thiol-specific, heat shock inducers results in thermal denaturation at 37 degrees C. Biochem 36(36):11002–11011

Shi Y, Kroeger PE, Morimoto RI (1995) The carboxyl-terminal transactivation domain of heat shock factor 1 is negatively regulated and stress responsive. Mol Cell Biol 15(8):4309–4318

Shi Y, Mosser DD, Morimoto RI (1998) Molecular chaperones as HSF1-specific transcriptional repressors. Genes Dev 12(5):654–666

Soncin F, Zhang X, Chu B et al (2003) Transcriptional activity and DNA binding of heat shock factor-1 involve phosphorylation on threonine 142 by CK2. Biochem Biophys Res Commun 303(2):700–706

Sorger PK, Nelson HC (1989) Trimerization of a yeast transcriptional activator via a coiled-coil motif. Cell 59(5):807–813

Sorger PK, Pelham HR (1987) Purification and characterization of a heat-shock element binding protein from yeast. EMBO J 6(10):3035–3041

Treuter E, Nover L, Ohme K, Scharf KD (1993) Promoter specificity and deletion analysis of three heat stress transcription factors of tomato. Mol Gen Genet 240(1):113–125

Voellmy R (1994) Transduction of the stress signal and mechanisms of transcriptional regulation of heat shock/stress protein gene expression in higher eukaryotes. Crit Rev Eukaryot Gene Expr 4(4):357–401

Vuister GW, Kim SJ, Wu C, Bax A (1994) NMR evidence for similarities between the DNA-binding regions of Drosophila melanogaster heat shock factor and the helix-turn-helix and HNF-3/forkhead families of transcription factors. Biochem 33(1):10–16

Westwood JT, Clos J, Wu C (1991) Stress-induced oligomerization and chromosomal relocalization of heat-shock factor. Nature 353(6347):822–827

Westwood JT, Wu C (1993) Activation of Drosophila heat shock factor: conformational change associated with a monomer-to-trimer transition. Mol Cell Biol 13(6):3481–3486

Wiederrecht G, Seto D, Parker CS (1988) Isolation of the gene encoding the S. cerevisiae heat shock transcription factor. Cell 54(6):841–853

Wiegant FA, Malyshev IY, Kleschyov AL, van Faassen E, Vanin AF (1999) Dinitrosyl iron complexes with thiol-containing ligands and S-nitroso-D, L-penicillamine as inductors of heat shock protein synthesis in H35 hepatoma cells. FEBS Lett 455(1–2):179–182

Wisniewski J, Orosz A, Allada R, Wu C (1996) The C-terminal region of Drosophila heat shock factor (HSF) contains a constitutively functional transactivation domain. Nucleic Acids Res 24(2):367–374

Wu C (1995) Heat shock transcription factors: structure and regulation. Annu Rev Cell Dev Biol 11:441–469

Zhang Y, Huang L, Zhang J, Moskophidis D, Mivechi NF (2002) Targeted disruption of hsf1 leads to lack of thermotolerance and defines tissue-specific regulation for stress-inducible Hsp molecular chaperones. J Cell Biochem 86(2):376–393

Zhao C, Hashiguchi A, Kondoh K, Du W, Hata J, Yamada T (2002) Exogenous expression of heat shock protein 90 kDa retards the cell cycle and impairs the heat shock response. Exp Cell Res 275(2):200–214

Zou J, Guo Y, Guettouche T, Smith DF, Voellmy R (1998) Repression of heat shock transcription factor HSF1 activation by HSP90 (HSP90 complex) that forms a stress-sensitive complex with HSF1. Cell 94(4):471–480

Zuo J, Baler R, Dahl G, Voellmy R (1994) Activation of the DNA-binding ability of human heat shock transcription factor 1 may involve the transition from an intramolecular to an intermolecular triple-stranded coiled-coil structure. Mol Cell Biol 14(11):7557–7568

Zuo J, Rungger D, Voellmy R (1995) Multiple layers of regulation of human heat shock transcription factor 1. Mol Cell Biol 15(8):4319–4330

Zhong M, Orosz A, Wu C (1998) Direct sensing of heat and oxidation by Drosophila heat shock transcription factor. Mol Cell 2(1):101–108

Chapter 5
HSP70 in the Immune Responses

Abstract HSP70 s play important roles in immune responses. The specific physiological context substantially influences the immune functions of HSP70. The first factor is the localization of HSP70; whether it is intracellular, on the cell surface, or in circulation. Intracellular HSP70 protects the cell and restricts cytokine production, whereas extracellular HSP70 stimulates cytokine production and labels cells for destruction. The second factor is the type of receptors on the target cells that bind HSP70. Signaling receptors, such as the toll-like receptor (TLR), confer to HSP70 the ability to activate cytokine production and stimulate the innate response, whereas scavenger receptors help HSP70 to deliver antigens to antigen-presenting cells and therefore stimulate an adaptive response. The third factor is the circumstances of synthesis and release of HSP70 from the cell. For example, in the case of microbial invasion, HSP70 s are involved in the formation of antigen-dependent immune memory, and in the case of different stresses in the formation of antigen-independent immune memory.

Keywords HSP70 • Immunity • Immune memory • Inflammation • Cytokines • NFkB

Humans have always been inclined to believe that they have a very special place in nature that is different from all other living species. Only humans invented the wheel and learned how to make fire. Only humans created great civilizations that changed the world in the ways that no other organism could do. Therefore for a long time it seemed natural that the human body lives and gets sick according to its own human laws. That is why until the 16th century diseases were studied using an anatomical approach, i.e., when people dissected other people.

However, physicians could not study the causes and pathogenesis of diseases simply by performing autopsies. Fortunately, help came from Darwin's theory of evolution. The theory proved that the major principles of normal development are the same for all living organisms. Subsequently, one smart pathophysiologist got the idea that diseases in different species also develop according to the same common laws. That means, he thought, that human diseases can be studied using animals and even bacteria. His name was Claude Bernard. Therefore, once again, after the homeostasis theory, we have to say "thank you, Claude Bernard!" Thanks to research on animals, cells and bacteria we have developed an understanding about the important role of HSPs in normal and damaged cells.

This chapter starts by examining medical aspects of HSP70, beginning with the role of HSP70 in immune responses.

5.1 Immune Response: Steps of Development

Figure 5.1 shows the main steps of development of innate and adaptive immunity. Macrophages are the first to detect pathogens in the body. Interacting with intracellular organisms—viruses and bacteria—macrophages produce proinflammatory cytokines, such as IL-12 and TNF-α, as well as chemokines (Janeway et al. 2005; Mantovani et al. 2004, 2006; Mantovani 2006). These chemokines attract natural killers, neutrophils and T-cells into the focus of inflammation (Boehm et al. 1997; Sharma 2010). IL-12 and TNF-α affect the activity of natural killer cells and macrophages, and an increase in the secretion of IFN-γ by these cells is observed. IFN-γ further stimulates macrophages to produce IL-12 and TNF-α, and this cytokine also enhances the phagocytic and bactericidal properties of macrophages (Trost et al. 2009; Nelson 2001).

When interacting with the extracellular parasites—fungi and worms—macrophages produce anti-inflammatory cytokines, such as IL-10 and chemokines (Mantovani et al. 2004; Mantovani 2006; Mantovani et al. 2006; Martinez et al. 2006). These chemokines attract T lymphocytes, eosinophils and basophils, which produce IL-4 and IL-13. IL-4 and IL-13 stimulate macrophages to further produce

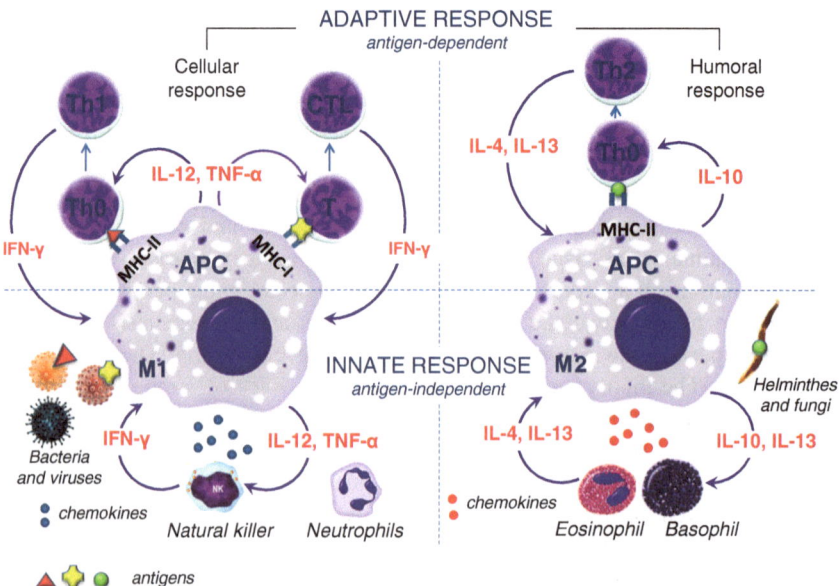

Fig. 5.1 The main steps of development of innate and adaptive immunity

IL-10 (Kaplan 2001; Falcone et al. 2000). IL-10 reduces the production of pro-inflammatory cytokines (D'Andrea et al. 1993), reactive oxygen species and NO (Hu et al. 1995) and, therefore, reduces the bactericidal properties of macrophages.

This step marks the development of the innate immune response.

To successfully remove the pathogen, the body launches an adaptive immune response. How does this happen?

Antigens of intracellular microbes, the pro-inflammatory phenotype of macrophages and its cytokines TNF-α, IL-12 and IFN-γ potentiate the development of Th0 cells into Th1 cells. The response of Th1 cells helps to neutralize viruses, bacteria and cancer cells (Mantovani et al. 2004). Antigens of extracellular parasites, the the anti-inflammatory phenotype of macrophages and its cytokines IL-10 and IL-4 potentiate the development of Th0 cells into Th2 cells (Sieling et al. 1993). The Th2 humoral response neutralizes extracellular bacteria, parasites and toxins.

5.2 HSP70 Protects Immune Cells, but "A Spoon is Dear When Lunch Time is Near"

At the point of an inflammation, macrophages are considered to be in unfavorable conditions. To survive, macrophages start synthesizing protective HSP70. This HSP70 synthesis is activated by cytokines and NO, which are produced by macrophages themselves. Thus, the first obvious function of HSP70 in the immune response is to protect immune cells from damaging inflammatory factors.

The example of the role of HSP70 in inflammation will show an enormous difference between the laws of mathematics and biology. No one has ever doubted the law of mathematics saying that changing the order of the addends does not change the sum; $2 + 3$ and $3 + 2$ will always equal five. However, it works differently in biology (Fig. 5.2). For instance, it was shown that prior accumulation of HSP70 protects cells from inflammation. If you "change the order of addends" and activate

Fig. 5.2 The heat shock paradox: prior accumulation of HSP70 protects cells from inflammation, however activation of the HSP70 synthesis after the initial stages of inflammation kills the cells

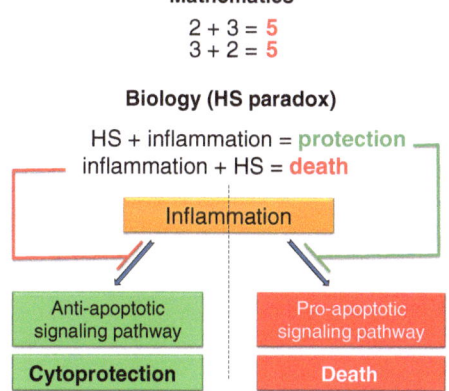

the HSP70 synthesis after the initial stages of inflammation, the cells will die! This situation is called the heat shock paradox. Inflammation simultaneously induces both anti-apoptotic and pro-apoptotic pathways. If HSP70 synthesis is activated before inflammation, HSP70 will block apoptosis, whereas if HSP70 is activated after the start of inflammation, it will enhance apoptosis. Thus, the activation of HSP70 synthesis must be timely. Indeed, "a spoon is dear when lunch time is near".

5.3 Role of HSP70 in the Development of the Innate Response, or the Duplicity of the Chaperone

Activation of HSP70 synthesis and its accumulation in macrophages is important for the development of the innate response (Fig. 5.3). HSP70 s have been shown to block the activation of pro-inflammatory cytokine genes through inhibition of a cytokine transcription factor—nuclear factor-kB (NF-kB) (Cahill et al. 1996, 1997; Housby et al. 1999; Ianaro et al. 2001; Ding et al. 2001; Xie et al. 2002a, b). As mentioned, HSP70 synthesis in macrophages can be activated by these same cytokines. This means that the accumulation of HSP70 in the cell closes the negative feedback. This regulatory mechanism restricts excessive production of pro-inflammatory cytokines, which may kill neighboring normal cells and thereby does not allow the situation of "an innocent passer-by syndrome" during inflammation (Yoo et al. 2000; Zugel et al. 1999; Lindquist and Craig 1988).

It was long thought that HSP70 is a typical intracellular protein. However, since the late 1980 s, more exceptions have appeared. For example, researchers began

Fig. 5.3 The role of HSP70 in the control of the innate response: HSP70 s block the activation of NF-kB and thus restrict excessive production of pro-inflammatory cytokines

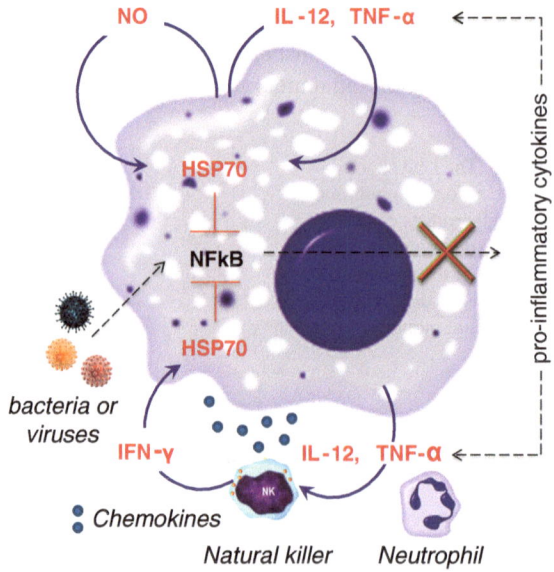

finding HSP70 molecules on the surface of normal, infected and tumor cells, and also in the blood stream (Hirsh et al. 2006; Egan and Carding 2000; Multhoff et al. 1997; Di Cesare et al. 1992; Ferrarini et al. 1992; Poccia et al. 1996; Sapozhnikov et al. 2002; Hashiguchi et al. 2001; Carding and Egan 2000; Belles et al. 1999; Johnson and Fleshner 2006; Tytell et al. 1986; Hightower and Guidon 1989; Mabula and Calderwood 2006a, b)

Significant efforts have been made to answer two key questions: how do these proteins exit cells? What is the function of extracellular HSP70?

First of all, the conditions under which HSP70 s are found outside the cell have been analyzed, as well as what types of cells can release HSP70.

It appeared that a variety of cells, such as neurons, monocytes and macrophages, B cells and tumor cells (Robinson et al. 2005; Clayton et al. 2005; Davies et al. 2006) can release HSP70 under very different circumstances: (i) in severe heat shock or under the action of toxic substances (Mabula and Calderwood 2006a, b; Todryk et al. 1999); (ii) necrosis (Wewers 2004; Mabula and Calderwood 2006a, b); (iii) various diseases (Wright et al. 2000; Pockley 2002; Pockley et al. 2003); (iv) aging (Terry et al. 2004); and (v) under various types of stress conditions and physical exercise (Campisi and Fleshner 2003; Pockley 2002; Fleshner et al. 2006; Fleshner and Johnson 2005; Pittet et al. 2002).

An analysis of this list shows that in most cases cells release HSP70 in pathological conditions, wherever there is cell damage and necrosis. This immediately leads to an obvious conclusion: the appearance of HSP70 in the extracellular space may be related with cell lysis and the discharge of intracellular contents (Srivastava 2003; Calderwood 2005).

Is necrosis the only mechanism for the release of HSPs from cells? The second part of the list clearly indicates that it certainly is not! Physiological stress and physical activity are not accompanied by cell necrosis. Experiments have also confirmed that glial, (Guzhova et al. 2001), tumor (Gastpar et al. 2005), mononuclear (Lancaster and Febbraio 2005) and B cells (Clayton et al. 2005) can release HSP70 in the absence of detectable cell death. So there must be other mechanisms involved.

The studies of Lancaster and Febbraio (2005) and others (Gastpar et al. 2005; Clayton et al. 2005; Lancaster and Febbraio 2005; Bausero et al. 2005) showed that exosomes provide the main pathway for the physiological release of HSP72 (Fig. 5.4). Furthermore, Johnson and Fleshner (2006) established that the increased output of HSP70-containing exosomes during physiological stress is a result of the activated sympathetic nervous system and norepinephrine release. Norepinephrine, through the activation of adrenergic receptors, increases intracellular Ca^{2+} (Guarino et al. 1996), which stimulates the release of HSP70-containing exosomes (Savina et al. 2003).

Thus, the lysis of cells in pathological situations or hormone-dependent release of exosomes in normal conditions represents the main mechanisms of HSP70 release from cells.

What consequences may result from the emergence of HSP70 on the cell surface or in circulation? The structural similarity between the microbial and human HSP70 s (Karlin and Brocchieri 1998) was a good tip for those who reflected on

Fig. 5.4 Exosomes provide
the main pathway for the
physiological release of
HSP70 from cells

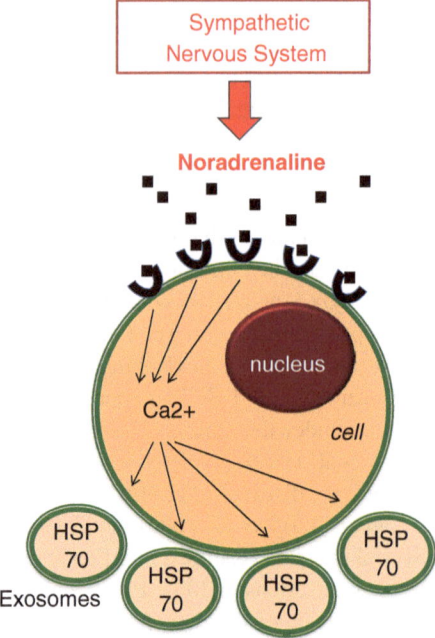

the role of extracellular HSP70! The first who used this "tip" was Dr. Shinnick
(Shinnick et al. 1996). He was the first to suggest the similarities and then, along
with others, showed that extracellular HSP70 may trigger a strong immune
response (Zügel and Kaufmann 1999; Shinnick 1991).

Different immune cells were shown to be able to recognize HSP70 on the cell
surface. For example, on the surface of tumor cells, HSP70 s are recognized by
natural killer cells, dendritic cells and cytotoxic T-cells. This recognition enables
natural killer cells to kill virus-infected and tumor cells (Multhoff 2002; Multhoff
et al. 1997, 1999; Moser et al. 2002; Gross et al. 2003a, b; Gastpar et al. 2004),
and dendritic (Chen et al. 2009) and cytotoxic T-cells are activated (Srivastava
2002a, b).

Thus, it became clear that endogenous HSP70 exposed on the outer surface of
cell membranes contributes to the development of innate and adaptive immune
responses.

In parallel, scientists were focusing on free HSP70 found in circulation. Here,
too, interesting and important details have emerged. It turned out that circulating
HSP70 can bind to the outer membrane of macrophages, dendritic cells and T-cells
(Srivastava 2002a, b; Gross et al. 2003a, b; Figueiredo et al. 2009). Subsequently,
HSP70 was shown to bind microbial products and modulate toll-like receptor[1]

---

[1] Toll-like receptors (TLRs) are a class of proteins that play a key role in the innate immune
system. They are membrane-spanning receptors that recognize structurally conserved molecules
derived from microbes.

(TLR)-dependent signaling mechanisms (Osterloh et al. 2007). These studies suggested that circulating free HSP70 may be involved in the immune response.

Starting in 1993, this hypothesis has received strong support. Since then, multiple studies have reported that HSP70 preparations isolated from mammals (Asea et al. 2000a, b; Basu et al. 2000; Somersan et al. 2001; Panjwani et al. 2002) and recombinant human HSP70 (Vabulas et al. 2002; Asea et al. 2000a, b; Asea et al. 2002; Dybdahl et al. 2002) can activate the innate response. These studies have demonstrated that free extracellular HSP70 stimulates production of the proinflammatory cytokines TNF-α (Campisi and Fleshner 2003; Asea et al. 2000a, b), interleukin 1 (IL-1) (Campisi and Fleshner 2003; Asea et al. 2000a, b), IL-6 (Campisi and Fleshner 2003; Asea et al. 2000a, b), and IL-12 (Multhoff et al. 1999; Breloer et al. 2001), NO (Campisi and Fleshner 2003; Panjwani et al. 2002) and chemokines in monocytes, macrophages and dendritic cells (Asea et al. 2000a, b; Srivastava 2002a, b; Panjwani et al. 2002; Lehner et al. 2000a, b). Furthermore, the ability of HSP70 to stimulate the production of innate response mediators did not depend on whether the peptide (antigen) was bound to HSP70 (Asea et al. 2000a, b).

It was also reported that free HSP70 induces the maturation of dendritic cells, because it has been shown that an increase in the number of major histocompatibility complex[2] (MHC) class I and II molecules occurs, and also an increase in the levels of co-stimulatory molecules Cluster of Differentiation 80 (CD80), CD86 and CD40 (Basu et al. 2000; Somersan et al. 2001; Wang et al. 2002).

In science, establishment of an interesting scientific phenomenon, especially a medically relevant one, has always been a powerful incentive and motivation to search for mechanisms of this phenomenon. This was certainly the case with HSP70 s role in the immune response! Studies have determined that the cytokine effects of HSP70 are mediated through CD40 and CD14/TLR (2 and 4) signaling pathways of NFκB and mitogen-activated protein kinases (MAPK) activation (Vabulas et al. 2002; Asea et al. 2000a, b; Asea et al. 2002; Basu et al. 2000; Somersan et al. 2001; Srivastava 2002a, b; Gross et al. 2003a, b; Asea et al. 2000a, b; Quintana et al. 2004; Quintana and Cohen 2005).

HSP70 also interacts with CD91 in antigen-presenting cells and other types of immune cells (Basu et al. 2001; Binder et al. 2000a, b)

Thus, the role of HSP70 in the innate response depends on its localization in a polar way. Such "duplicity" of a chaperone at the molecular level is explained by HSP70 having different "targets": intracellular HSP70 aids the maintenance of protein homeostasis and inhibits NFkB, a cytokine transcription factor, whereas extracellular HSP70 attracts killer cells and activates receptors activating NFkB.

---

[2] Major histocompatibility complex (MHC) is a cell surface molecule encoded by a large gene family in all vertebrates. MHC molecules mediate interactions of immune cells, with other leukocytes or body cells. In humans, MHC is also called human leukocyte antigen (HLA).

5.4 HSP70 as an Informer of the Immune System that Detects Damages to the Body

Naturally the question arises: what is it all for? What is the biological rationale for HSP70 being involved in the immune response? According to Cao (Chen and Cao 2010) and Ireland (Williams and Ireland 2008) HSP70, which is released at the site of tissue damage, enters into the circulation system and performs cytokine-like functions. Consequently, HSP70 can serve as a signal of damage or alert the host immune system. This hypothesis was further confirmed when an increase in circulating HSP70 was found in different diseases: renal disease (Wright et al. 2000), hypertension and atherosclerosis (Pockley et al. 2003).

This would all be very well; however, the level of HSP70 in the blood can increase in response to physiological stress or physical exercise (Fleshner et al. 2004; Lancaster et al. 2004; Campisi and Fleshner 2003; Campisi et al. 2003; Fleshner et al. 2003; Walsh et al. 2001; Febbraio et al. 2002), even when there is no tissue damage or medical condition. This understanding leads us to an important conclusion, which does not counter, but expands the concept of the "alarm" signals: the stress-induced release of HSP70 is a feature of a normal stress response (Fleshner et al. 2006), and HSP70 plays a signalling role to inform the innate immunity about the presence of stress.

5.5 The Role of HSP70 in the Development of an Adaptive Response, or the Antigenic Escort

Immune effects of extracellular HSP70 are not restricted to the innate response. HSP70 was found to play an important role in the adaptive response (Srivastava 2002a, b), namely, in the activation and maturation of dendritic cells in the antigen presentation and activation of T-cells (Srivastava 2002a, b; Lehner et al. 2004; Srivastava and Heike 1991; Srivastava 2002a, b; Srivastava et al. 1994; Li et al. 2002; Calderwood et al. 2007; Pockley et al. 2008).

The role of HSP70 in the activation and maturation of dendritic cells is defined by the fact that HSP70 stimulates migration of dendritic cells (Binder et al. 2000a, b) and increases the expression of surface CD40, CD83, CD86 and MHC class II molecules (Basu et al. 2000; Singh-Jasuja et al. 2000; Asea et al. 2002; Noessner et al. 2002).

However, HSP70 plays the most significant role in antigen presentation. HSP70 may be involved in antigen presentation at three key stages of this process (Javid et al. 2007; Noessner et al. 2002): (1) during formation of the complex with a peptide antigen; (2) during antigenic peptide delivery to antigen-presenting cells and of the antigen transfer into the cell; and (3) during intracellular chaperoning of antigens to the MHC class I molecules.

The HSP70-antigen complex can form both outside and inside cells, and be released during necrosis (MacAry et al. 2004; Noessner et al. 2002; Singh-Jasuja et al. 2000; Grossmann et al. 2004; Suto and Srivastava 1995). In performing the chaperone functions, HSP70 binds the antigenic peptide via the substrate domain (Zhu et al. 1996). In situations with low ATP levels, HSP70 exists in the ADP-bound form, and thus strongly retains the protein substrate (see Chap. 1). Now it becomes clear that the shortage of ATP in damaged cells and low blood levels of ATP will contribute to the formation of the "HSP70-antigen" complex.

How important is the binding of the antigen to HSP70? Extremely important! Several studies have shown that "HSP70-antigen" complexes significantly poten-tiate the activation of T-cells (Li et al. 2002; Bendz et al. 2007) and anti-tumor immunity (Javid et al. 2004) compared with scenarios with the antigen alone or without a HSP.

On the surface of antigen presenting cells the HSP70-antigen complex can bind to either signaling receptors such as TLR2, TLR4 and CD91, or to scavenger receptors (SR), such as LOX-1, SREC-1, FEEL-1/CLEVER-1 and CD91 (Arnold-Schild et al. 1999; Singh-Jasuja et al. 2000; Theriault et al. 2005; Delneste et al. 2002a, b; Theriault et al. 2006). Binding of the complex to signaling receptors results in the activation of cytokine production, i.e., the induction of an innate response. The interaction with the scavenger receptors or CD91 results in recep-tor-mediated endocytosis of the complex (Li et al. 2002; Basu et al. 2001; Todryk et al. 1999; Delneste et al. 2002a, b; Chu and Pizzo 1993; Binder et al. 2000a, b, 2004; Takemoto et al. 2005) and the antigen entry into the cell.

When an antigenic peptide enters the cell, HSP70 continues to "chaperone" the antigen and transports it to the endoplasmic reticulum (Srivastava 2002a, b; Martin et al. 2003). HSP70 enters into the cell in the ADP conformation, which allows HSP70 to securely hold the protein antigen, thereby ensuring that the peptide is not released. The antigen-presenting cell, in contrast to the extracel-lular space and circulation, has significant ATP levels and nucleotide-exchange factors. Therefore, HSP70 readily exchanges ADP for ATP and, in the ATP con-formation, releases the bound peptide antigen to facilitate entry into the endo-plasmic reticulum.

In the endoplasmic reticulum, the preparation for the immune antigen presenta-tion continues. There, other heat shock proteins, namely HSP90, bind the deliv-ered antigenic peptides and transport them to class I MHC to present the antigen on the surface of antigen-presenting cell for interaction with CD8[+] receptors on T lymphocytes (Li et al. 2002; Ishii et al. 1999; Srivastava 2008) and activation of adaptive immunity.

It may appear strange and unlikely that one molecule performs so many immune functions. Multiplicity of receptors for HSP70 on the surface of immune cells explains the immunological multifunctionality of HSP70. TLR, for exam-ple, confers to HSP70 the possibility to activate cytokine production in an NFkB-dependent way and to stimulate the innate response, whereas SR allows HSP70 to deliver antigenic peptides into antigen-presenting cells (Li et al. 2002) and to stimulate the adaptive immune response (Fig. 5.5).

Fig. 5.5 Extracellular
HSP70 can play a molecular
immune switch role, toggling
between innate and adaptive
responses

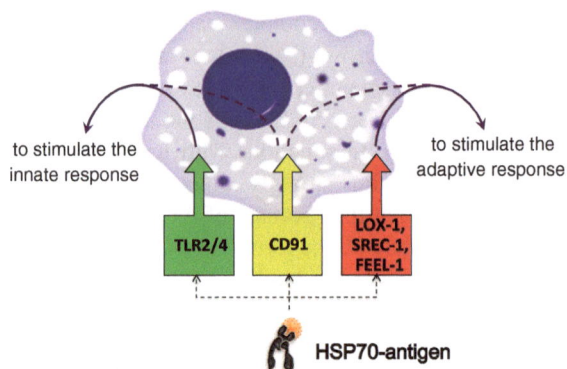

Thus, extracellular HSP70 can play a molecular immune switch role, toggling between innate and adaptive responses (Wang et al. 2005).

At the same time, it appeared quite interesting that the ability of HSP70 to stimulate innate and adaptive responses involves different domains of the HSP70 molecule (Lehner et al. 2004; Wang et al. 2002, 2005; Lehner et al. 2000a, b). The C-terminal domain stimulates production of chemokines, IL-12, TNF-α, NO and binds to CD14, TLR4, and CD40 on antigen-presenting cells, i.e., stimulates the innate response (Lehner et al. 2004; Wang et al. 2001, 2002, 2005; Lehner et al. 2000a, b). The substrate domain binds the antigenic peptide and thus stimulates the antigen presentation and the adaptive response (MacAry et al. 2004). Finally, the N-terminal ATPase domain, through the ADP/ATP exchange, regulates both binding and release of the antigenic peptide, which may be related to stimulation of both the innate and adaptive responses.

5.6 HSP70, Immune Memory Cells, or "I Remember Everything That did Not Happen to Me!"

Perhaps the most important consequence of immune responses to a specific pathogenic antigen is the formation of immunological memory. Immunological memory is the ability of the immune system to respond more quickly and effectively to the antigen with which the body has already met. Such memory is ensured by pre-existing antigen-specific clones of B- and T-memory cells. Therefore, when the antigen re-enters the body, memory cells will recognize the antigen and will form an immune response more quickly and effectively.

Among scientists, the thesis of antigen-specificity of the immune memory became a sort of "unquestionable truth" since Karl Landsteiner (1868–1943, Austria) predicted the existence of such a memory almost 100 years ago, and it has never been questioned. Until 2010, Dr. Thomas Lehner from the National Institutes of Health did not question this dogma either. However, when he was studying the role of HSP70 on the surface of dendritic cells some doubt crept into his mind (Wang et al. 2010). In this study, HSP70 appeared on dendritic cells after heat or oxidative stress. The emergence of HSP70 resulted in the activation of NFkB and a subsequent increase in IL-15 on the surface of dendritic cells (Fig. 5.6). These IL-15 molecules bound to their receptors on the surface of CD4(+)T-cells to induce the appearance of CD40 ligands on the surface of T-cells and proliferation of T-cells. The CD40 ligand on CD4(+)T cells, in turn, reactivated CD40 on dendritic cells, inducing maturation of dendritic cells and an increase in the expression of IL-15. Thus a positive feedback mechanism was formed, which stimulated the formation of CD62L(+) T memory cells from CD4 (+) T cells.

Thus it was first shown that T memory cells can be formed independently of the antigen and MHC class II molecules, and that stress-induced HSP70 on the surface of dendritic cells plays the leading role in this process!

However, until that time "the classic of the immune genre" assumed that without the antigen and the MHC-II-TCR interaction neither activation nor proliferation of T cells happened! Open any textbook on immunology and you will see that the formation of T memory cells is considered only in the context of antigen-specificity.

But why does the immune system need to form antigen-independent memory and to remember what stress represents? In humans, high temperature and stress accompany virtually all specific infections, so the formation of an immunological memory of stress through an emerging pool of antigen-independent T memory cells may help the body to cope faster with any new infection. This phenomenon could be defined as the phenomenon of cross-immunological memory. In this phenomenon, stress-induced HSP70 allows the immune system to respond to a new infection as if the immune system somehow knew and remembered the infection, though there has never been any previous contact.

Fig. 5.6 T memory cells can be formed independently of the antigen and MHC class II molecules, and HSP70 on the surface of dendritic cells plays the leading role in this process

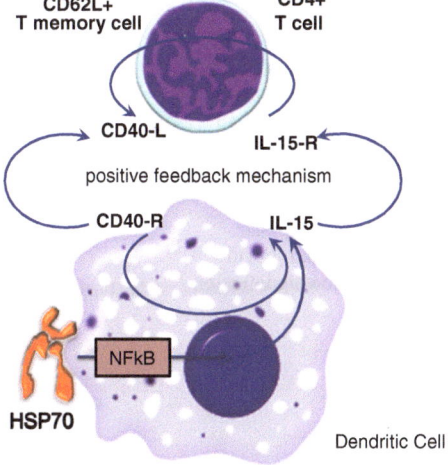

5.7 HSP70 in the Termination of Inflammation

A rapid immune response to penetrating pathogens is certainly important for the host defense against infection. However, termination of this response is equally important to prevent damage to tissue by toxic mediators of inflammation. Activated macrophages and neutrophils release the largest amount of these mediators. Therefore, timely removal of these cells is a prerequisite for a proper shutdown of inflammation. It was found that one of the mechanisms for removing the inflammatory cells is associated with HSP60 and HSP70. Basically, this is how it works.

In the conditions of excessive inflammation, macrophages and neutrophils increase HSP60 and HSP70 synthesis, and expose them to their surface (Fig. 5.7). HSP molecules in mammals resemble microbial counterparts, and therefore phagocytes decorated with HSP60 and HSP70 are recognized by the special γδT killer cells as target cells (O'Brien et al. 1992). These γδT cells recognize the "death" label (HSPs), and kill the overactivated immune cells that have been destroying the pathogenic bacteria (Hirsh et al. 2006). Just like in the criminal world, where the assassin kills the killer after the murder.

In contrast to the criminal world, however, the removal of the long-activated immune cells, serves to protect the host organs from being damaged by aggressive phagocytes (Hirsh et al. 2004, 2006; Moore et al. 2000; Chung et al. 2006; Saunders et al. 1998; Tam et al. 2001). Indeed, it has been shown that γδT cells that recognize HSP60 and HSP70 play an important role in protecting the body from secondary damage, such as immune responses to peritoneal sepsis (Hirsh et al. 2004, 2006), experimental *Listeria monocytogenes* (Kimura et al. 1998), *Mycobacteria tuberculosis* (Beagley et al. 1993) or *Plasmodium malariae* (Tsuji et al. 1994).

Thus HSP70 and HSP60, by attracting γδT cells, play an important role in removing excessively activated phagocytes, and consequently, in the natural termination of inflammation and recovery following infection (Born et al. 1999; Egan and Carding 2000; Carding and Egan 2000; Belles et al. 1999).

Fig. 5.7 The role of HSPs in the termination of inflammation: HSP70 and HSP60, by attracting γδT cells remove excessively activated phagocytes, and consequently, terminate inflammation and provide recovery following infection

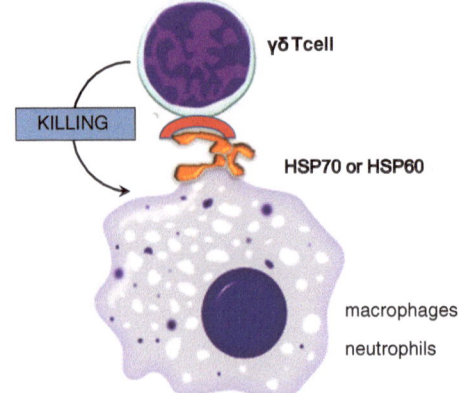

5.8   Some Doubts About the Role of HSP70 in the Immune Response: Dr Gao's Fly in the Ointment

Immunnomodulating properties of extracellular HSP70 have been so impressive that it caused enormous interest and a huge increase in the number of articles on this subject followed. A large number of beautiful hypotheses were put forward. However, in early the 2000s Dr. Gao had sounded a warning that was in disagreement with the general consensus (Gao and Tsan 2003a, b). He and his colleagues had found that many of the HSP70 immune effects in vitro may be due to lipopolysaccharide[3] contamination of exogenous HSP70 preparations. As often happens; "The great tragedy of science − the slaying of a beautiful hypothesis by an ugly fact" (Huxley 1870).

All cytokine-like effects in vitro have been carefully re-checked on HSP preparations, thoroughly cleaned of contamination. Together with Gao (Gao and Tsan 2003a, b; Tsan and Gao 2004; 2007) Wallin et al. (2002), Bausinger et al. (2002), Reed et al. (2003), Osterloh et al. (2007) and Bendz et al. (2008) all indicated that the cytokine effects of HSPs are unlikely to be correct.

Thus, any in vitro studies should be interpreted with care. Nonetheless, the data obtained in vivo have been so convincing that we can in good faith repeat once again a well-founded conclusion that endogenous extracellular HSP70 contributes to the development and termination of immune responses.

It is important to keep in mind that the immune effects of HSP70 are highly dependent on several factors.

The first factor is the localization of HSP70 to either intracellular, cell surface, or circulation. Intracellular HSP70 protects the cell and restricts cytokine production, whereas extracellular HSP70, in contrast, stimulates cytokine production and labels cells for destruction.

The second factor is the type of receptors on the target cells that bind HSP70. Signaling receptors, such as TLR, confer to HSP70 the ability to activate cytokine production and stimulate the innate response, whereas scavenger receptors help HSP70 to deliver antigens into antigen-presenting cells and stimulate an adaptive response.

The third factor relates to the synthesis and release of HSP70 from the cell. For example, in the case of microbial invasion, HSP70 is involved in the formation of antigen-dependent immune memory, and in the case of different stresses HSP70 is involved in the formation of antigen-independent immune memory.

Thus, the specific physiological context substantially influences the immune functions of HSP70. This is a warning to all researchers: refrain from any global statements about the role of HSP70 in the immune response in vivo.

---

[3] Lipopolysaccharides are large molecules consisting of a lipid and a polysaccharide joined by a covalent bond. They are found in the outer membrane of Gram-negative bacteria, act as endotoxins and elicit strong immune responses in animals.

References

Arnold-Schild D, Hanau D, Spehner D et al (1999) Cutting edge: receptor-mediated endo-cytosis of heat shock proteins by professional antigen-presenting cells. J Immunol 162:3757–3760

Asea A, Kabingu E, Stevenson MA, Calderwood SK (2000a) HSP70 peptide-bearing and pep-tide-negative preparations act as chaperokines. Cell Stress Chaperones 5:425–431

Asea A, Kraeft SK, Kurt-Jones EA et al (2000b) HSP70 stimulates cytokine production through a CD-14-dependent pathway, demonstrating its dual role as a chaperone and cytokine. Nat Med 6:435–442

Asea A, Rehli M, Kabingu E et al (2002) Novel signal transduction pathway utilized by extracel-lular HSP70: role of Toll-like receptor (TLR) 2 and TLR4. J Biol Chem 277:15028–15034

Basu S, Binder RJ, Suto R, Anderson KM, Srivastava PK (2000) Necrotic but not apoptotic cell death releases heat shock proteins, which deliver a partial maturation signal to dendritic cells and activate the NF-κB pathway. Int Immunol 12:1539–1546

Basu S, Binder RJ, Ramalingam T, Srivastava PK (2001) CD91 is a common receptor for heat shock proteins gp96, HSP90, HSP70, and calreticulin. Immunity 14:303–313

Bausero MA, Gastpar R, Multhoff G, Asea A (2005) Alternative mechanism by which IFN-{γ} enhances tumor recognition: active release of heat shock protein 72. J Immunol 175:2900–2912

Bausinger H, Lipsker D, Ziylan U et al (2002) Endotoxin-free heat shock protein 70 fails to induce APC activation. Eur J Immunol 32:3708–3713

Beagley KW, Fujihashi K, Black CA et al (1993) The Mycobacterium tuberculosis 71-kDa heat-shock protein induces proliferation and cytokine secretion by murine gut intraepithelial lym-phocytes. Eur J Immunol 23(8):2049–2052

Belles C, Kuhl A, Nosheny R, Carding SR (1999) Plasma membrane expression of heat shock protein 60 in vivo in response to infection. Infect Immun 67(8):4191–4200

Bendz H, Ruhland SC, Pandya MJ et al (2007) Human heat shock protein 70 enhances tumor antigen presentation through complex formation and intracellular antigen delivery without innate immune signaling. J Biol Chem 282:31688–31702

Bendz H, Marincek BC, Momburg F et al (2008) Calcium signaling in dendritic cells by human or mycobacterial HSP70 is caused by contamination and is not required for HSP70-mediated enhancement of cross-presentation. J Biol Chem 283:26477–26483

Binder RJ, Anderson KM, Basu S, Srivastava PK (2000a) Cutting edge: heat shock protein gp96 induces maturation and migration of CD11c + cells in vivo. J Immunol 165(11):6029–6035

Binder RJ, Han DK, Srivastava PK (2000b) CD91: a receptor for heat shock protein gp96. Nat Immunol 1(2):151–155

Binder RJ, Vatner R, Srivastava P (2004) The heat-shock protein receptors: some answers and more questions. Tissue Antigens 64(4):442–451

Boehm U, Klamp T, Groot M, Howard JC (1997) Cellular responses to interferon-gamma. Annu Rev Immunol 15:749–795

Born W, Cady C, Jones-Carson J, Mukasa A, Lahn M, O'Brien R (1999) Immunoregulatory functions of gamma delta T cells. Adv Immunol 71:77–144

Breloer M, Dorner B, Moré SH, Roderian T, Fleischer B, von Bonin A (2001) Heat shock pro-teins as "danger signals": eukaryotic HSP60 enhances and accelerates antigen-specific IFN-γ production in T cells. Eur J Immunol 31:2051–2059

Cahill CM, Waterman WR, Xie Y, Auron PE, Calderwood SK (1996) Transcriptional repression of the prointerleukin 1ß gene by heat shock factor 1. J Biol Chem 271:24874–24879

Cahill CM, Lin HS, Price BD, Bruce JL, Calderwood SK (1997) Potential role of heat shock transcription factor in the expression of inflammatory cytokines. Adv Exp Med Biol 400B:625–630

Calderwood SK (2005) Regulatory interfaces between the stress protein response and other gene expression programs in the cell. Methods 35(2):139–148

Calderwood SK, Mambula SS, Gray PJ Jr, Theriault JR (2007) Extracellular heat shock proteins in cell signaling. FEBS Lett 581:3689–3694

Campisi J, Fleshner M (2003) The role of extracellular HSP72 in acute stress-induced potentiation of innate immunity in physically active rats. J Appl Physiol 94:43–52

Campisi J, Leem TH, Fleshner M (2003) Stress-induced extracellular HSP72 is a functionally significant danger signal to the immune system. Cell Stress Chaperones 8:272–286

Carding SR, Egan PJ (2000) The importance of gamma delta T cells in the resolution of pathogen-induced inflammatory immune responses. Immunol Rev 173:98–108

Chen T, Cao X (2010) Stress for maintaining memory: HSP70 as a mobile messenger for innate and adaptive immunity. Eur J Immunol 40:1541–1544

Chen T, Guo J, Han C, Yang M, Cao X (2009) Heat shock protein 70, released from heat-stressed tumor cells, initiates antitumor immunity by inducing tumor cell chemokine production and activating dendritic cells via TLR4 pathway. J Immunol 182:1449–1459

Chu CT, Pizzo SV (1993) Receptor-mediated antigen delivery into macrophages. Complexing antigen to alpha 2-macroglobulin enhances presentation to T cells. J Immunol 150(1):48–58

Chung CS, Watkins L, Funches A, Lomas-Neira J, Cioffi WG, Ayala A (2006) Deficiency of gammadelta T lymphocytes contributes to mortality and immunosuppression in sepsis. Am J Physiol Regul Integr Comp Physiol 291(5):R1338–R1343

Clayton A, Turkes A, Navabi H, Mason MD, Tabi Z (2005) Induction of heat shock proteins in B-cell exosomes. J Cell Sci 118:3631–3638

D'Andrea A, Aste-Amezaga M, Valiante NM, Ma X, Kubin M, Trinchieri G (1993) Interleukin 10 (IL-10) inhibits human lymphocyte interferon gamma-production by suppressing natural killer cell stimulatory factor/IL-12 synthesis in accessory cells. J Exp Med 178(3):1041–1048

Davies EL, Bacelar MM, Marshall MJ et al (2006) Heat shock proteins form part of a danger signal cascade in response to lipopolysaccharide and GroEL. Clin Exp Immunol 145:183–189

Delneste Y, Charbonnier P, Herbault N et al (2002a) Interferon-gamma switches monocyte differentiation from dendritic cells to macrophages. Blood 101(1):143–150

Delneste Y, Magistrelli G, Gauchat J et al (2002b) Involvement of LOX-1indendriticcell-mediatedantigen cross-presentation. Immunity 17:353–362

Di Cesare S, Poccia F, Mastino A, Colizzi V (1992) Surface expressed heat-shock proteins by stressed or human immunodeficiency virus (HIV)-infected lymphoid cells represent the target for antibody-dependent cellular cytotoxicity. Immunology 76(2):341–343

Ding XZ, Fernandez-Prada CM, Bhattacharjee AK, Hoover DL (2001) Over-expression of HSP-70 inhibits bacterial lipopolysaccharide-induced production of cytokines in human monocyte-derived macrophages. Cytokine 16:210–219

Dybdahl B, Wahba A, Lien E et al (2002) Inflammatory response after open heart surgery: release of heat-shock protein 70 and signaling through Toll-like receptor-4. Circulation 105:685–690

Egan PJ, Carding SR (2000) Downmodulation of the inflammatory response to bacterial infection by gammadelta T cells cytotoxic for activated macrophages. J Exp Med 191(12):2145–2158

Falcone FH, Haas H, Gibbs BF (2000) The human basophil: a new appreciation of its role in immune responses. Blood 96(13):4028–4038

Febbraio MA, Ott P, Nielsen HB et al (2002) Exercise induces hepatosplanchnic release of heat shock protein 72 in humans. J Physiol 544:957–962

Ferrarini M, Heltai S, Zocchi MR, Rugarli C (1992) Unusual expression and localization of heat-shock proteins in human tumor cells. Int J Cancer 51(4):613–619

Figueiredo C, Wittmann M, Wang D et al (2009) Heat shock protein 70 (HSP70) induces cytotoxicity of T-helper cells. Blood 113:3008–3016

Fleshner M, Johnson JD (2005) Exogenous extra-cellular heat shock protein 72: releasing signal(s) and function. Int J Hyperthermia 21:457–471

Fleshner M, Campisi J, Johnson JD (2003) Can exercise stress facilitate innate immunity? A functional role for stress-induced extracellular HSP72. Exerc Immunol Rev 9:6–24

Fleshner M, Campisi J, Amiri L, Diamond DM (2004) Cat exposure induces both intra- and extracellular HSP72: the role of adrenal hormones. Psychoneuroendocrinology 29(9):1142–1152

Fleshner M, Sharkey CM, Nickerson M, Johnson JD (2006) Endogenous extracellular HSP72 release is an adaptive feature of the acute stress response. In: Ader R, Felton DL, Cohen N (eds) Psychoneuroimmunology, vol 2. Academic Press, San Diego, pp 1013–1014

Gao B, Tsan M-F (2003a) Recombinant human heat shock protein 60 does not induce the release of tumor necrosis factor α from murine macrophages. J Biol Chem 278:22523–2252

Gao B, Tsan M-F (2003b) Endotoxin contamination in recombinant human HSP70 prepara- tion is responsible for the induction of TNFα release by murine macrophages. J Biol Chem 278:174–179

Gastpar R, Gross C, Rossbacher L, Ellwart J, Riegger J, Multhoff G (2004) The cell surface- localized heat shock protein 70 epitope TKD induces migration and cytolytic activity selec- tively in human NK cells. J Immunol 172:972–980

Gastpar R, Gehrmann M, Bausero MA et al (2005) Heat shock protein 70 surface-positive tumor exosomes stimulate migratory and cytolytic activity of natural killer cells. Cancer Res 65:5238–5247

Gross C, Koelch W, DeMaio A, Arispe N, Multhoff G (2003a) Cell surface-bound heat shock protein 70 (HSP70) mediates perforin-indepen- dent apoptosis by specific binding and uptake of granzyme B. J Biol Chem 278:41173–41181

Gross C, Schmidt-Wolf IG, Nagaraj S et al (2003b) Heat shock protein 70-reactivity is associ- ated with increased cell surface density of CD94/CD56 on primary natural killer cells. Cell Stress Chaperones 8(4):348–360

Grossmann ME, Madden BJ, Gao F et al (2004) Proteomics shows HSP70 does not bind peptide sequences indiscriminately in vivo. Exp Cell Res 297:108–117

Guarino RD, Perez DM, Piascik MT (1996) Recent advances in the molecular pharmacology of the α 1-adrenergic receptors. Cell Signal 8:323–333

Guzhova I, Kislyakova K, Moskaliova O et al (2001) In vitro studies show that HSP70 can be released by glia and that exogenous HSP70 can enhance neuronal stress tolerance. Brain Res 914:66–73

Hashiguchi N, Ogura H, Tanaka H et al (2001) Enhanced expression of heat shock proteins in activated polymorphonuclear leukocytes in patients with sepsis. J Trauma 51(6):1104–1109

Hightower LE, Guidon PT (1989) Selective release from cultured mam-malian cells of heat- shock (stress) proteins that resemble glia-axon transfer proteins. J Cell Physiol 138:257–266

Hirsh M, Dyugovskaya L, Kaplan V, Krausz MM (2004) Response of lung gammadelta T cells to experimental sepsis in mice. Immunology 112(1):153–160

Hirsh MI, Hashiguchi N, Chen Y, Yip L, Junger WG (2006) Surface expression of HSP72 by LPS-stimulated neutrophils facilitates gammadelta T cell-mediated killing. Eur J Immunol 36(3):712–721

Housby JN, Cahill CM, Chu B et al (1999) Non-steroidal anti-inflammatory drugs inhibit the expression of cytokines and induce HSP70 in human monocytes. Cytokine 11:347–358

Hu S, Sheng WS, Peterson PK, Chao CC (1995) Differential regulation by cytokines of produc- tion of nitric oxide by human astrocytes. Glia 15(4):491–494

Huxley TH (1870) Biogenesis and Abiogenesis. In: Collected Essays, Vol 8, p 229

Ianaro A, Ialenti A, Maffia P, Pisano B, Di Rosa M (2001) HSF1/HSP72 pathway as an endog- enous anti-inflammatory system. FEBS Lett 499:239–244

Ishii T, Udono H, Yamano T et al (1999) Isolation of MHC class 1-restricted tumor antigen pep- tide and its precursors associated with heat shock proteins HSP70, HSP90, and gp96. J Immunol 162:1303–1309

Janeway CA, Travers P, Walport M, Shlomchik MJ, Capra JD (2005) Immunobiology: the immune system in health and disease. Garland Science Publishing, USA

Javid B, MacAry PA, Oehlmann W, Singh M, Lehner PJ (2004) Peptides complexed with the protein HSP70 generate efficient human cytolytic T-lymphocyte responses. Biochem Soc Trans 32(Pt 4):622–625

Javid B, MacAry PA, Lehner PJ (2007) Structure and function: heat shock proteins and adaptive immunity. J Immunol 179(4):2035–2040

Johnson JD, Fleshner M (2006) Releasing signals, secretory pathways, and immune function of endogenous extracellular heat shock protein 72. J Leukoc Bio 79:425–434

Kaplan AP (2001) Chemokines, chemokine receptors and allergy. Int Arch Allergy Immunol 124(4):423–431

Karlin S, Brocchieri L (1998) Heat shock protein 70 family: multiple sequence comparisons, function, and evolution. J Mol Evol 47(5):565–577

Kimura Y, Yamada K, Sakai T et al (1998) The regulatory role of heat shock protein 70-reactive CD4 + T cells during rat listeriosis. Int Immunol 10(2):117–130

Lancaster GI, Febbraio MA (2005) Exosome-dependent trafficking of HSP70: a novel secretory pathway for cellular stress proteins. J Biol Chem 280:23349–23355

Lancaster GI, Gleeson M, Jeukendrup AE et al (2004) Leukocyte heat shock protein expression before and after intensified training. Int J Sports Med 25(7):522–527

Lehner T, Bergmeier LA, Wang Y, Tao L, Sing M, Spallek R, van der Zee R (2000a) Heat shock proteins generate ß-chemokines which function as innate adjuvants enhancing adaptive immunity. Eur J Immunol 30:594–603

Lehner T, Mitchell E, Bergmeier L et al (2000b) The role of gammadelta T cells in generating antiviral factors and beta-chemokines in protection against mucosal simian immunodeficiency virus infection. Eur J Immunol 30(8):2245–2256

Lehner T, Wang Y, Whittall T, McGowan E, Kelly CG, Singh M (2004) Functional domains of HSP70 stimulate generation of cytokines and chemokines, maturation of dendritic cells and adjuvanticity. Biochem Soc Trans 32:629–632

Li Z, Menoret A, Srivastava P (2002) Roles of heat-shock proteins in antigen presentation and cross-presentation. Curr Opin Immunol 14:45–51

Lindquist S, Craig EA (1988) The heat-shock proteins. Annu Rev Genet 22:631–677

Mabula SS, Calderwood SK (2006a) Heat induced release of HSP70 from prostate carcinoma cells involves both active secretion and passive release from necrotic cells. Int J Hyperthermia 22:575–585

Mabula SS, Calderwood SK (2006b) Heat shock protein 70 is secreted from tumor cells by a nonclassical pathway involving lysosomal endosomes. J Immunol 177:7849–7857

MacAry PA, Javid B, Floto RA, Smith KG, Oehlmann W, Singh M, Lehner PJ (2004) HSP70 peptide binding mutants separate antigen delivery from dendritic cell stimulation. Immunity 20:95–106

Mantovani A (2006) Macrophage diversity and polarization.: in vivo veritas. Blood 108(2):408–409

Mantovani A, Sica A, Sozzani S, Allavena P, Vecchi A, Locati M (2004) The chemokine system in diverse forms of macrophage activation and polarization. Trends Immunol 25:677–686

Mantovani A, Sica A, Locati M (2006) New vistas on macrophage differentiation and activation. Eur J Immunol 37(1):14–16

Martin CA, Carsons SE, Kowalewski R (2003) Aberrant extracellular and dendritic cell (DC) surface expression of heat shock protein (HSP)70 in the rheumatoid joint: possible mechanisms of HSP/DC-mediated cross-priming. J Immunol 171:5736–5742

Martinez FO, Gordon S, Locati M, Mantovani A (2006) Transcriptional profiling of the human monocyte-to-macrophage differentiation and polarization: new molecules and patterns of gene expression. J Immunol 177:7303–7311

Moore TA, Moore BB, Newstead MW, Standiford TJ (2000) Gamma delta-T cells are critical for survival and early proinflammatory cytokine gene expression during murine Klebsiella pneumonia. J Immunol 165(5):2643–2650

Moser C, Schmidbauer C, Gürtler U et al (2002) Inhibition of tumor growth in mice with severe combined immunodeficiency is mediated by heat shock protein 70 (HSP70)-peptide-activated, CD94 positive natural killer cells. Cell Stress Chaperones 7(4):365–373

Multhoff G (2002) Activation of natural killer cells by heat shock protein 70. Int J Hyperthermia 18(6):576–585

Multhoff G, Botzler C, Jennen L, Schmidt J, Ellwart J, Issels R (1997) Heat shock protein 72 on tumor cells: a recognition structure for natural killer cells. J Immunol 158:4341–4350

Multhoff G, Mizzen L, Winchester CC et al (1999) Heat shock protein 70 (HSP70) stimulates proliferation and cytolytic activity of NK cells. Exp Hematol 27:1627–1636

Nelson S (2001) Novel nonantibiotic therapies for pneumonia cytokines and host defence. Chest 119(2):419S–425S

Noessner E, Gastpar R, Milani V et al (2002) Tumor-derived heat shock protein 70 peptide complexes are cross-presented by human dendritic cells. J Immunol 169:5424–5432

O'Brien RL, Fu YX, Cranfill R et al (1992) Heat shock protein HSP60-reactive gamma delta cells: a large, diversified T-lymphocyte subset with highly focused specificity. Proc Natl Acad Sci U S A 89(10):4348–4352

Osterloh A, Kalinke U, Weiss S, Fleischer B, Breloer M (2007) Synergistic and differential modulation of immune responses by HSP60 and LPS. J Biol Chem 282:4669–4680

Panjwani NN, Popova L, Srivastava PK (2002) Heat shock protein gp96 and HSP70 activate the release of nitric oxide by APCs. J Immunol 168:2997–3003

Pittet JF, Lee H, Morabito D, Howard MB, Welch WJ, Mackersie RC (2002) Serum levels of HSP 72 measured early after trauma correlate with survival. J Trauma 52:611–617

Poccia F, Piselli P, Vendetti S, Bach S, Amendola A, Placido R, Colizzi V (1996) Heat-shock protein expression on the membrane of T cells undergoing apoptosis. Immunology 88(1):6–12

Pockley AG (2002) Heat shock proteins, inflammation, and cardiovascular disease. Circulation 105:1012–1017

Pockley AG, Georgiades A, Thulin T, de Faire U, Frostegård J (2003) Serum heat shock protein 70 levels predict the development of atherosclerosis in subjects with established hypertension. Hypertension 42(3):235–238

Pockley AG, Muthana M, Calderwood SK (2008) The dual immunoregulatory roles of stress proteins. Trends Biochem Sci 33:71–79

Quintana FJ, Cohen IR (2005) Heat Shock Proteins Regulate Inflammation by Both Molecular and Network Cross-Reactivity. Cambridge University Press, Cambridge

Quintana FJ, Carmi P, Mor F, Cohen IR (2004) Inhibition of adjuvant-induced arthritis by DNA vaccination with the 70-kd or the 90-kd human heat-shock protein: immune cross-regulation with the 60-kd heat-shock protein. Arthritis Rheum 50:712–3720

Reed RC, Berwin B, Baker JP, Nicchitta CV (2003) GRP94/gp96 elicits ERK activation in murine macrophages: a role for endotoxin contamination in NFκB activation and nitric oxide production. J Biol Chem 278:31853–31860

Robinson MB, Tidwell JL, Gould T et al (2005) Extracellular heat shock protein 70: a critical com- ponent for motoneuron survival. J Neurosci 25:9735–9745

Sapozhnikov AM, Gusarova GA, Ponomarev ED, Telford WG (2002) Translocation of cytoplasmic HSP70 onto the surface of EL-4 cells during apoptosis. Cell Prolif 35(4):193–206

Saunders BM, Frank AA, Cooper AM, Orme IM (1998) Role of gamma delta T cells in immunopathology of pulmonary *Mycobacterium avium* infection in mice. Infect Immun 66(11):5508–5514

Savina A, Furlán M, Vidal M, Colombo MI (2003) Exosome release is regulated by a calcium-dependent mechanism in K562 cells. J Biol Chem 278:20083–20090

Sharma M (2010) Chemokines and their receptors: orchestrating a fine balance between health and disease. Crit Rev Biotechnol 30(1):1–22

Shinnick TM (1991) Heat shock proteins as antigens of bacterial and parasitic pathogens. Curr Top Microbiol Immunol 167:145–160

Shinnick TM, Coulson AF, Oftung F, Mustafa AS, Lundin KE, Meloen RH (1996) HLA-DR4-restricted T-cell epitopes from the mycobacterial 60,000 MW heat shock protein (HSP 60) do not map to the sequence homology regions with the human HSP 60. Immunology 87(3):421–427

Sieling PA, Abrams JS, Yamamura M et al (1993) Immunosuppressive roles for IL-10 and IL-4 in human infection. In vitro modulation of T cell responses in leprosy. J Immunol 150(12):5501–5510

Singh-Jasuja H, Toes RE, Spee P et al (2000) Cross-presentation of glycoprotein 96-associated antigens on major histocompatibility complex class I molecules requires receptor-mediated endocytosis. J Exp Med 191:1965–1974

Somersan S, Larsson M, Fonteneau JF, Basu S, Srivastava P, Bhardwaj N (2001) Primary tumor tissue lysates are enriched in heat shock proteins and induce the maturation of human dendritic cells. J Immunol 167:4844–4852

Srivastava PK (2002a) Interaction of heat shock proteins with peptides and antigen presenting cells: chaperoning of the innate and adaptive immune responses. Annu Rev Immunol 20:395–425

Srivastava PK (2002b) Roles of heat-shock proteins in innate and adaptive immunity. Nat Rev Immunol 2:185–194

Srivastava PK (2003) Hypothesis: controlled necrosis as a tool for immunotherapy of human cancer. Cancer Immun 18(3):4

Srivastava PK (2008) New jobs for ancient chaperones. Sci Am 299:50–55

Srivastava PK, Heike M (1991) Tumor-specific immunogenicity of stress-induced proteins: convergence of two evolutionary pathways of antigen presentation? Semin Immunol 3:57–64

Srivastava PK, Udono H, Blachere NE, Li Z (1994) Heat shock proteins transfer peptides during antigen processing and CTL priming. Immunogenetics 39:93–98

Suto R, Srivastava PK (1995) A mechanism for the specific immunogenicity of heat shock protein-chaperoned peptides. Science 269:1585–1588

Takemoto S, Nishikawa M, Takakura Y (2005) Pharmacokinetic and tissue distribution mechanism of mouse recombinant heat shock protein 70 in mice. Pharm Res 22(3):419–426

Tam S, King DP, Beaman BL (2001) Increase of gammadelta T lymphocytes in murine lungs occurs during recovery from pulmonary infection by *Nocardia asteroides*. Infect Immun 69(10):6165–6171

Terry DF, McCormick M, Andersen S et al (2004) Cardiovascular disease delay in centenarian offspring: role of heat shock proteins. Ann N Y Acad Sci 1019:502–505

Thériault JR, Mambula SS, Sawamura T, Stevenson MA, Calderwood SK (2005) Extracellular HSP70 binding to surface receptors present on antigen presenting cells and endothelial/epithelial cells. FEBS Lett 579:1951–1960

Thériault JR, Adachi H, Calderwood SK (2006) Role of scavenger receptors in the binding and internalization of heat shock protein 70. J Immunol 177:8604–8611

Todryk S, Melcher AA, Hardwick N et al (1999) Heat shock protein 70 induced during tumor cell killing induces Th1 cytokines and targets immature dendritic cell precursors to enhance antigen uptake. J Immunol 163:1398–1408

Trost M, English L, Lemieux S, Courcelles M, Desjardins M, Thibault P (2009) The phagosomal proteome in interferon-γ-activated macrophages. Immunity 30(1):143–154

Tsan M-F, Gao B (2004) Cytokine function of heat shock proteins. Am J Physiol Cell Physiol 286:C739–C744

Tsan M-F, Gao B (2007) Pathogen-associated molecular pattern contamination as putative endogenous ligands of toll-like receptors. J Endotoxin Res 13:6–14

Tsuji M, Mombaerts P, Lefrancois L, Nussenzweig RS, Zavala F, Tonegawa S (1994) Gamma delta T cells contribute to immunity against the liver stages of malaria in alpha beta T-cell-deficient mice. Proc Natl Acad Sci U S A 91(1):345–349

Tytell M, Greenberg SG, Lasek RJ (1986) Heat shock-like protein is transferred from glia to axon. Brain Res 363:161–164

Vabulas RM, Ahmad-Nejad P, Ghose S, Kirschning CJ, Issels RD, Wagner H (2002) HSP70 as endogenous stimulus of the Toll/interleukin-1 receptor signal pathway. J Biol Chem 277:15107–15112

Wallin RP, Lundqvist A, Moré SH, von Bonin A, Kiessling R, Ljunggren HG (2002) Heat-shock proteins as activators of the innate immune system. Trends Immunol 23:130–135

Walsh RC, Koukoulas I, Garnham A, Moseley PL, Hargreaves M, Febbraio MA (2001) Exercise increases serum HSP72 in humans. Cell Stress Chaperones 6:386–393

Wang Y, Kelly CG, Karttunen JT et al (2001) CD40 is a cellular receptor mediating mycobacterial heat shock protein 70 stimulation of CC-chemokines. Immunity 15:971–983

Wang Y, Kelly CG, Singh M et al (2002) Stimulation of Th-1 polarizing cytokines, C–C chemokines, maturation of dendritic cells, and adjuvant function by the peptide binding fragment of heat shock protein 70. J Immunol 169:2422–2429

Wang Y, Whittall T, McGowan E et al (2005) Identification of stimulating and inhibitory epitopes within the heat shock protein 70 molecule that modulate cytokine production and maturation of dendritic cells. J Immunol 174:3306–3316

Wang Y, Seidl T, Whittall T, Babaahmady K, Lehner T (2010) Stress-activated dendritic cells interact with CD4(1) T cells to elicit homeostatic memory. Eur J Immunol 40:1628–1638

Wewers MD (2004) IL-1beta: an endosomal exit. Proc Natl Acad Sci U S A 101:10241–10242

Williams JH, Ireland HE (2008) Sensing danger–HSP72 and HMGB1 as candidate signals. J Leukoc Biol 83:489–492

Wright BH, Corton JM, El-Nahas AM, Wood RF, Pockley AG (2000) Elevated levels of circulating heat shock protein 70 (HSP70) in peripheral and renal vascular disease. Heart Vessels 15(1):18–22

Xie Y, Chen C, Stevenson MA, Auron PE, Calderwood SK (2002a) Heat shock factor 1 represses transcription of the IL-1ß gene through physical interaction with the nuclear factor of interleukin 6. J Biol Chem 277:11802–11810

Xie Y, Chen C, Stevenson MA, Hume DA, Auron PE, Calderwood SK (2002b) NF-IL6 and HSF1 have mutually antagonistic effects on transcription in monocytic cells. Biochem Biophys Res Commun 291:1071–1080

Yoo CG, Lee S, Lee CT, Kim YW, Han SK, Shim YS (2000) Anti-inflammatory effect of heat shock protein induction is related to stabilization of I κ B α through preventing I κ B kinase activation in respiratory epithelial cells. J Immunol 164:5416–5423

Zhu W, Roma P, Pirillo A, Pellegatta F, Catapano AL (1996) Human endothelial cells exposed to oxidized LDL express HSP70 only when proliferating. Arterioscler Thromb Vasc Biol 16(9):1104–1111

Zügel U, Kaufmann SH (1999) Role of heat shock proteins in protection from and pathogenesis of infectious diseases. Clin Microbiol Rev 12(1):19–39

Chapter 6
HSP70 in Carcinogenesis

Abstract Depending on the location, HSP70 has different, often opposite effects on carcinogenesis. Intracellular HSP70 contributes to tumour development by: (1) supporting protein homeostasis in a tumour cell, thus protecting the cell from the adverse conditions of external inflammation; (2) contributing to the proliferation of tumour cells because HSP70 stabilizes cyclin D1; and (3) suppressing oncogene-induced apoptosis and the aging program. As a result, intracellular HSP70 creates the most favourable internal conditions for tumour growth. Membrane-associated and extracellular HSP70, in contrast, mainly aid the immune system to destroy the tumour. Extracellular HSP70 may participate in antigen-presentation of a tumour specific antigen and facilitate the development of anti-tumour adaptive responses. Extracellular HSP70 released from tumour cells, can influence the immune system even in the absence of an antigenic peptide. Natural killer cells can recognize HSP70 located on the tumour cell membrane as a tumour-specific structure. Along with natural killer cells, T memory cells can also recognize and kill HSP70-positive tumour cells.

Keywords HSP70 • Immunity • Tumor • Apoptosis • Cell senescence • P53

In the previous chapter I discussed the role of HSP70 in the development of an immune response. This role can be briefly summarized by three main concepts:

The first concept: intracellular HSP70 protects the cell and limits cytokine production, whereas extracellular HSP70, in contrast, stimulates cytokine production and labels cells for destruction.

The second concept: signalling receptors, such as TLR, confer to HSP70 the ability to activate cytokine production and stimulate the innate response, whereas scavenger receptors, such as SR, help HSP70 to deliver antigens to antigen-presenting cells and stimulate an adaptive response.

The third concept: in microbial invasion, HSP70 is involved in the formation of antigen-dependent immune memory, whereas under various stress conditions, HSP70 is involved in the formation of antigen-independent immune memory.

So what happens when the synthesis of these proteins is disrupted?

Some effects are obvious (Fig. 6.1). If macrophages lose the ability to synthesize protective HSP70, they die, and this considerably weakens the immune

I. Malyshev, *Immunity, Tumors and Aging: The Role of HSP70*,
SpringerBriefs in Biochemistry and Molecular Biology,
DOI: 10.1007/978-94-007-5943-5_6, © The Author(s) 2013

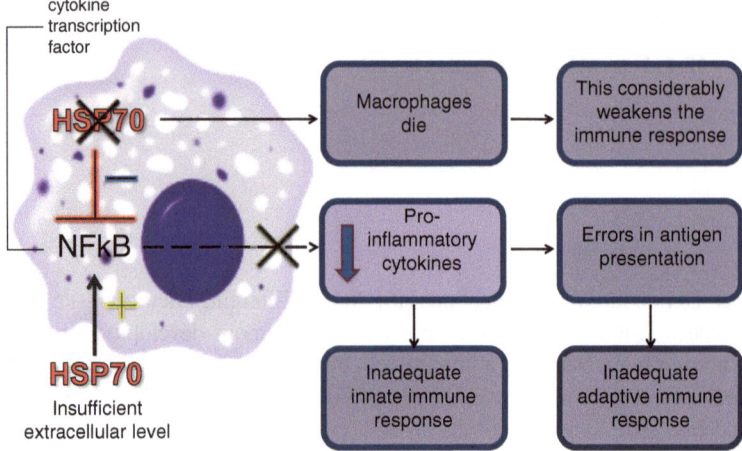

Fig. 6.1 What happens when the synthesis of HSP70 is disrupted

response. An insufficient extracellular level of HSP may lead to a decrease in cytokine production, errors in antigen presentation and, therefore, to an inadequate immune response.

However, there is another, perhaps more dramatic consequence of dysfunctional HSP70 synthesis and function, namely the inefficiency of immune surveillance and the role of HSP70 in carcinogenesis. Evaluating the role of HSP70 in carcinogenesis is not an easy task. There are more than two hundred types of tumours, each with its own characteristics (Tang et al. 2005) (Hahn and Weinberg 2002) and HSP70 is involved in many stages of tumour growth: proliferation, blocking apoptotic and senescence programs, invasion and metastasis (Romanucci et al. 2008). Here we will talk about the key immunological aspect.

It is obvious that ineffective immunity leads to a significant increase in cancer cases and premature death of patients. Failure of the immune system and modern medicine to resist the onslaught of transformed cells poses a real threat to the evolution of Homo Sapiens! By the way, a hypothesis on the extinction of dinosaurs proposes that dinosaurs lacked a developed immune system and died from cancer when the environmental conditions changed. Clearly, understanding and developing drugs that combat cancer represents a significant challenge that is being met by a large body of scientists. Since HSP70 plays an important role in regulating the immune response, this protein has attracted significant research efforts.

First of all, it was discovered that large amounts of HSP70 are present in various types of human tumours, such as breast cancer, endometrial cancer, lung cancer, prostate cancer and other tumour types (Jäättelä 1995; Vargas-Roig et al. 1997; Costa et al.1997; Nanbu et al. 1998; Ciocca et al. 1993; Santarosa et al. 1997).

The mechanism of abnormal activation of HSP70 synthesis in tumours is not fully understood. The main hypothesis suggests that HSP70 synthesis is induced by the appearance of abnormal denatured proteins in the cell (Voellmy 2004). The appearance of these proteins may be caused by low levels of glucose, pH and oxygen in the micro-environment of tumour cells, or gene mutations present in genes such as p53.

The increase in the synthesis of HSP70 in tumour cells often correlates with an increase in proliferation, metastasis and poor survival of patients (Jäättelä 1999; Ciocca and Calderwood 2005; Calderwood et al. 2006), and is often a poor prognostic sign (Jäättelä 1995; Garrido et al. 1998).

When the tumour develops, the HSP70 levels increase inside the cell, on the surface of cell membrane and in the extracellular space. Depending on the location, HSP70 has different, often opposite effects on carcinogenesis: (1) intracellular HSP70 contributes to tumour development by protecting tumour cells from adverse environmental conditions and by suppressing anti-tumour mechanisms, such as apoptosis and senescence whereas; (2) membrane-associated and extracellular HSP70 mainly help the immune system to destroy the tumour.

6.1 Intracellular HSP70 in Tumours: Rejuvenation Elixir or the Road to Hell is Paved by Good Intentions

When macrophages recognize tumour cells, they recognize these cells as potential danger and begin producing cytokines, NO and other inflammatory mediators to kill the tumour cells in a similar manner to the way bacteria are destroyed (Fig. 6.2). However, unlike bacteria, tumour cells respond to pro-inflammatory mediators like macrophages, by increasing the synthesis of their own HSP70. Tumour cells generally have high levels of intracellular HSP70. These HSP70 molecules, in "good faith" and "not knowing what they are doing", protect cancer cells from the cytotoxic action of macrophages. We can characterize the function of HSP70 in tumour cells with the phrase, "the road to hell is paved by good intentions."

The protection of tumour cells from the pro-inflammatory micro-environment by HSP70 is very unfortunate, but this is only half the trouble! Back in the mid-1990s, it was suspected that intracellular HSP70 directly promoted proliferation of tumour cells (Jäättelä 1995; Volloch and Sherman 1999; Seo et al. 1996). For example, in breast cancer an increase in HSP70 production leads to the stabilization of cyclin D1 (Diehl et al. 2003), which shortens the G0/G1 phase of the cell cycle and significantly accelerates cell division (Barnes et al. 2001). However that

Fig. 6.2 Tumour cells respond to macrophage pro-inflammatory mediators by increasing the synthesis of their own HSP70, which protects tumour cells

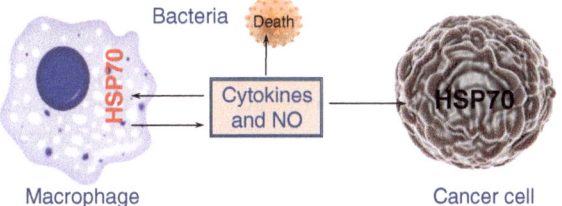

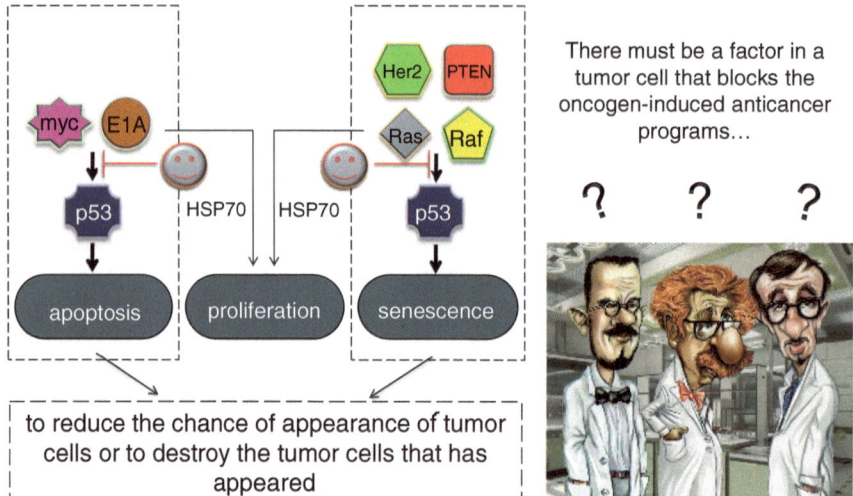

Fig. 6.3 Intracellular HSP70s block key anti-tumour programs in the cell, such as apoptosis and senescence

was not all! It turned out that intracellular HSP70 blocked key anti-tumour programs in the cell, such as apoptosis and senescence (Fig. 6.3).

In normal and tumour cells both apoptosis, or cell death, and senescence, or limiting the number of cell divisions, pursue the same goal: to reduce the appearance of tumour cells or destroy tumour cells that have appeared. Now imagine how unexpected and surprising it was when it was discovered that it was precisely the oncogenes that could activate apoptosis in tumour cells, as myc or E1A do (Pelengaris et al. 2000; Prendergast 1999; Perez-Sala and Rebollo 1999; White 1998; Blyth et al. 2000) or to trigger senescence, as Ras, Her-2, PTEN, Raf, and other oncogenes do (Ferbeyre et al. 2002; Trost et al. 2005; Chen et al. 2005; Sebastian and Johnson 2006; Benanti and Galloway 2004; Mason et al. 2004; Peeper et al. 2001; Chen et al. 2005; Olsen et al. 2002; Zhu et al. 1998). Interestingly, the oncogenes trigger both these anti-oncogene programs through activation of p53 (Chen et al. 2005; Nilsson and Cleveland 2003; Eischen et al. 1999; Yu et al. 1997).

So an intriguing question arises: whether the oncogene is an oncogene? What is a true oncogene in a cell? At the same time, no one has countered the concept that oncogenes can accelerate the proliferation of tumour cells. The only way out of this paradox was to conclude that the oncogene simultaneously triggers two programs, proliferation and apoptosis/senescence. For a normal cell it is quite appropriate, because this protects the cell from spontaneous malignancy during the action of normal growth factors. However, tumour cells apparently bypass this protection mechanism, therefore indicating that a factor in tumour cells is responsible for blocking oncogene-induced anticancer programs. So researchers began to look for this factor or mechanism.

Initially it appeared as if a quick answer had been found, because it was known that mutations in p53 were detected in many tumours, and this explained why p53-dependent apoptosis and senescence do not take place in cancer cells. Another mechanism for suppressing p53-dependent apoptosis/senescence in the process of malignant transformation could be associated with increases in the synthesis of the p53 transcription suppressor proteins Snail and Twist (Lee et al. 2009; Ansieau et al. 2008). However, it was found that a large number of tumours developed with a fully functionally preserved p53-dependent pathway. This means that there must be an alternative suppression factor for p53-dependent cell death. This factor was found to be HSP70. It "turns on a green light" in cells to suppress apoptosis and senescence, that is, this signal gives rise to cells with immortality and accelerated proliferation.

HSP70 showed its true, anti-tumour "personality" in experiments with HSP70 over-production and removal of this protein from cells. Dr. Guzhova from Russia was the first to demonstrate that HSP70 can suppress apoptosis stimulated by the myc oncogene (Afanasyeva et al. 2007). Other experiments were conducted by the Sherman (Sherman 2010) and Wei groups (Wei et al. 1995). These research teams showed that the removal of HSP70 from tumour cells resulted in the rapid development of apoptosis (Wei et al. 1995; Nylandsted et al. 2000a, b; Li et al. 2000; Gurbuxani et al. 2003;) and senescence (Sherman 2010). In both cases, there was strong activation of p53. Furthermore, when p53 induces the cell cycle inhibitor, gene p21, the senescence program is launched, and the cell cycle and division processes are arrested (Sherman 2010). However, if the induction of p21 is blocked, for example by myc, apoptosis is activated (Nilsson and Cleveland 2003; Eischen et al. 1999). At the same time, the removal of HSP70 did not affect the viability of normal cells (Nylandsted et al. 2000a, b; Gabai et al. 2000; Dix et al. 1996). This meant that tumour cells, in contrast to normal ones, cannot survive unless they activate the synthesis of HSP70. This observation is analogous to the fable describing where the needle of Koschei the Immortal was hidden! Koschei the Immortal is an evil character from Russian folk tales. Koschei cannot be killed by conventional means targeting his body. His soul is hidden separate from his body inside a needle, which is in an egg, which is in a duck, which is in a hare, which is in an iron chest, which is buried under a green oak tree, which is on the island of Buyan, in the ocean (Afanasyev 1873). As long as his soul is safe, he cannot die.

Thus, it was concluded that HSP70 inhibits the p53-dependent pathways of apoptosis and senescence in tumour cells.

Before discussing how HSP70 can block p53-dependent apoptosis in tumour cells, it necessary to say, that p53 can activate both the mitochondria- and receptor-dependent pathways of apoptosis.

p53 can trigger mitochondria-dependent apoptosis because this protein activates the proapoptotic members Bax and Bak, and inhibits the antiapoptotic members of the Bcl-2 family (Green and Kroemer 2009; Vaseva and Moll 2009; Leu et al. 2004; Mihara et al. 2003; Wu et al. 1997). As a result, Bax and Bak incorporate into the mitochondrial membrane and form pores, through which proapoptotic factors may exit. p53 can stimulate receptor-dependent apoptosis by activating death receptor genes (Müller et al. 1998; Bouvard et al. 2000; Bennett et al. 1998).

HSP70 directly binds to p53 (Nihei et al. 1993; Ehrhart et al. 1988) and maintains this protein in the inactive state. In addition, HSP70 can block p53-induced mitochondria- and receptor-dependent pathways of apoptosis. I have already discussed how HSP70 can do this (see Chap. 3).

An increase in HSP70 has been shown to lead to an increase in the level of antiapoptotic Bcl-2 (Kelly et al. 2002) and a reduction in the level of proapoptotic Bax (Stankiewicz et al. 2005). In this case, HSP70 blocks the possibility of Bax to incorporate into the outer mitochondrial membrane, thus preventing an increase in the permeability of mitochondrial membranes and the release of cytochrome c and AIF (Stankiewicz et al. 2005). In addition, HSP70 can directly block the exit of cytochrome c (Matsumori et al. 2006; Lee et al. 2004; Tsuchiya et al. 2003), proapoptotic protein Smac/DIABLO (Jiang et al. 2005) and Apaf-1 from mito-chondria (Matsumori et al. 2006; Beere et al. 2000; Saleh et al. 2000). HSP70 can bind to Apaf-1 thus preventing the attraction of procaspase 9 to the apoptosome (Saleh et al. 2000). HSP70 can also directly inhibit caspase-9 (Beere et al. 2000). HSP70 can reduce the number of death receptors through inhibition of JNK (Park et al. 2001; Lee et al. 2005). In addition, HSP70 can bind to death receptors DR4 and DR5 thus inhibiting transmission of the apoptotic signal through these recep-tors (Guo et al. 2005). HSP70 can limit apoptosis, when caspase activation has already occurred. For example, HSP70 can restrict both activation of phospholi-pase A2 and changes in nuclear morphology (Jäättelä et al. 1998). HSP70 can also prevent activation of caspase-independent apoptotic pathways (Creagh et al. 2000; Ravagnan et al. 2001), through the binding to AIF and EndoG and by blocking the translocation of proapoptotic factors into the nucleus (Sun et al. 2006; Ruchalski et al. 2006; Ravagnan et al. 2001; Gurbuxani et al. 2003; Matsumori et al. 2005; Kalinowska et al. 2005). Thus, HSP70 can inhibit the development of p53-depend-ent antitumour apoptosis (Steel et al. 2004; Stankiewicz et al. 2005; Ravagnan et al. 2001) by directly binding to p53, and through the inhibition of the active p53-induced mitochondria- and receptor-dependent apoptotic pathways.

Cell senescence is the second important anti-tumour cell program, which limits the number of cell divisions (Hayflick 1979; McCormick and Maher 1988). Replicative senescence, to some extent, ensures that a normal cell, which accumu-lates various potentially pro-oncogenic mutations during its life (Vogelstein and Kinzler 1993.) does not pass them on to their daughter cells, thus causing tumour progression (Wright and Shay 2001; Campisi 2005). Initially it was assumed that replicative senescence is the result of telomere[1] shortening. This has been con-firmed. However, another mechanism was found (Fig. 6.4). Replicative senescence may be triggered by p53, which for this purpose increases the expression of two endogenous inhibitors of the cell cycle, p16 and p21 (Braig and Schmitt 2006; Garbe et al. 2007; Li et al. 2007).

Thus oncogenes, such as Ras, activate a program of aging through several mechanisms that involve p53. For example, these oncogenes may activate the

---

[1] A telomere is a region of repetitive nucleotide sequences at each end of a chromosome, which protects the end of the chromosome from deterioration or from fusion with neighboring chromo-somes. Over time, due to each cell division, the telomere ends become shorter.

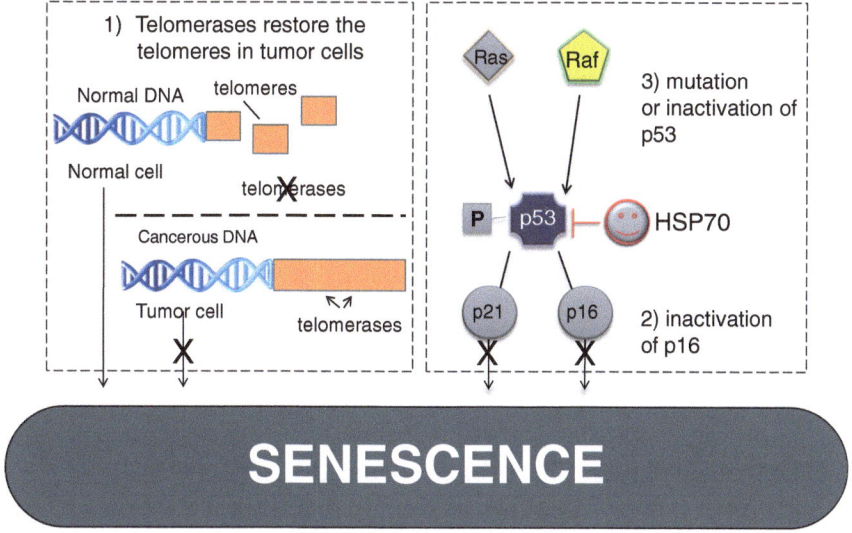

Fig. 6.4   HSP70 inhibits the p53-dependent pathways of senescence in tumour cells

DNA damage response ATM kinase, and ATM kinase in turn phosphorylates and stabilizes p53 (Di Micco et al. 2006). Another mechanism involves the induction of p19ARF, which inhibits the p53 ubiquitin ligase Hdm2, also leading to the stabilization of p53 (Bihani et al. 2004). This results in the accumulation and activation of p53.

In addition to activating the p53-dependent pathway, oncogenes, such as Ras or Raf, can activate aging with the help of p53-independent mechanisms. The main route involves MAP kinases and ends by the activation of extracellular-signal-regulated kinases (ERK)[2] (Yaswen and Campisi 2007).

Unfortunately, tumour cells do not undergo replicative senescence and become "immortalized" (Hanahan and Weinberg 2000). This dictates that tumour cells acquire mechanisms for avoiding replicative aging. The first of these mechanisms is related to the fact that telomerases, which restore the length of telomeres, are activated in tumour cells (Shay and Bacchetti. 1997; Kim et al. 1994) The second is connected with the inactivation of p16, either through mutation or by methylation of the promoter (Matsuda 2008; Schwabe and Lubbert 2007). The p16 protein is a cyclin-dependent kinase (CDK) inhibitor that decelerates the cell cycle by inactivating the CDKs that phosphorylate the retinoblastoma protein (pRb). The third mechanism is associated with a mutation or inactivation of p53.

---

[2] ERKs are involved in functions including the regulation of meiosis, mitosis and postmitotic functions in differentiated cells. Many different stimuli, including growth factors, cytokines, virus infection, ligands for G protein-coupled receptors, transforming agents, and carcinogens, activate the ERK pathway.

We have already mentioned that HSP70 can inactivate p53, thus likely suppressing the development of the aging program in tumour cells, and therefore providing tumour cells with immortality.

In general, the following hypothesis describes the role of intracellular HSP70 accumulation in tumour cells.

First, HSP70 through their chaperone functions support protein homeostasis in a tumour cell, thus protecting it from the adverse conditions of external inflammation. Second, HSP70 contributes to the proliferation of tumour cells due to the stabilization of cyclin D1. Third, HSP70 suppresses oncogene-induced apoptosis and the aging program, without affecting the ability of oncogenes to accelerate proliferation (Sherman 2010). As a result, intracellular HSP70 creates the most favourable internal conditions for tumour growth.

6.2 Extracellular and Membrane-Bound HSP70 in Tumours, or an Apology to the Host Immune System for its Intracellular Counterparts

In addition to intracellular localization, HSP70 can be found on the plasma membrane of malignantly transformed cells (Shin et al. 2003; Multhoff et al. 1995; Gross et al. 2003; Schmitt et al. 2007) and in the extracellular environment (Triantafilou and Triantafilou 2004). Screening of more than 1,000 tumour biopsies showed that 50–70 % of human carcinomas contain HSP70 integrated into the membrane, whereas normal tissues did not contain HSP70 on their surface (Multhoff et al. 1995; Multhoff 2007; Gehrmann et al. 2003).

There are two alternative views describing the role of HSP70 that has been integrated into the membrane of tumour cells and extracellular HSP70 that has been released from a tumour cell.

In the first point of view, it is assumed that the membrane-bound HSP70 aids in maintaining the stability of tumour cell membranes and thus may protect the tumour cells from the adverse surrounding environment (Horvath and Vigh 2010; Horvath et al. 2008). This view is supported by clinical studies that show that the appearance of HSP70 on the surface of tumour cells correlates with a decrease in the survival rate of patients with squamous cell carcinomas of the lung and lower rectal carcinomas (Pfister et al. 2007). However, other clinical studies have shown opposing results. In osteosarcoma and renal cancers, an increase in the expression of HSP70 was associated with an improved prognosis (Santarosa et al. 1997; Trieb et al. 1998). Therefore, another alternative point of view has been proposed and is possibly more justified. According to this proposal, extracellular and membrane HSP70 molecules facilitate the recognition of tumour cells and stimulate the antitumour response.

HSP70 released from tumour cells into the extracellular space and into circulation may carry intracellular tumour antigens (Fig. 6.5). The events could then unfold as Srivastava imagined them; we have already discussed these events in the

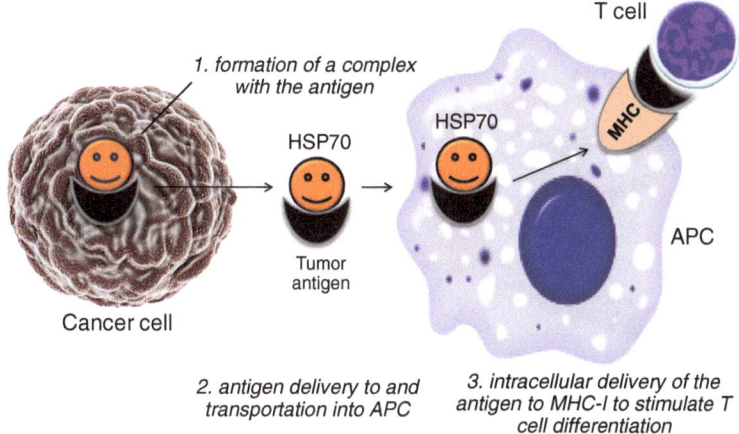

Fig. 6.5 HSP70 released from tumour cells may carry intracellular tumour antigens and then HSP70 may participate in antigen-presentation facilitating the development of anti-tumour adaptive responses

previous chapter. Basically, extracellular HSP70 may participate in antigen-presentation at three key steps of this process (Srivastava 2002): (1) formation of a complex with a peptide antigen; (2) antigenic peptide delivery to antigen-presenting cells and transport of the antigen into the cell; and (3) intracellular chaperoning of the antigen to the MHC class I molecules. These mechanisms certainly facilitate the development of anti-tumour adaptive responses.

According to Srivastava (Srivastava et al. 1998), extracellular HSP70 can influence the immune system even in the absence of an antigenic peptide (Fig. 6.6). It has been shown that together with proinflammatory cytokines IL-2 and IL-15, extracellular HSP70 can stimulate natural killers (Zeng et al. 2006; Multhoff et al. 1997), thus leading to the induction of antitumour immunity (Multhoff et al. 1997, 1999, 2001; Moser et al. 2002; Multhoff 2002; Gross et al. 2003). In this case, natural killers release granzyme B and induce apoptosis in tumour cells (Gross et al. 2003).

It was shown that natural killers can also recognize HSP70 located on the surface of tumour cell membranes as tumour-specific structures. The recognition happens because of the specific interaction of the CD94 receptor on the surface of natural killer cells and the peptide TKD-sequence located within the C-terminal part of HSP70 (Gross et al. 2003). This part of HSP70 extends into the extracellular environment of tumour cells and can be readily recognized (Multhoff et al. 2001). HSP70 on the membrane surface can also be recognized by the γδT memory cells, which we discussed in the previous chapter. That is why, along with natural killers, T memory cells can also kill HSP70-positive tumour cells (Laad et al. 1999).

It is interesting that HSP70 is integrated into the membrane only in tumour cells (Vega et al. 2008; Gehrmann et al. 2008), whereas HSP70 is bound to the membrane of normal antigen-presenting cells through a receptor. Perhaps that is why natural killer cells and T memory cells attack tumour cells, rather than

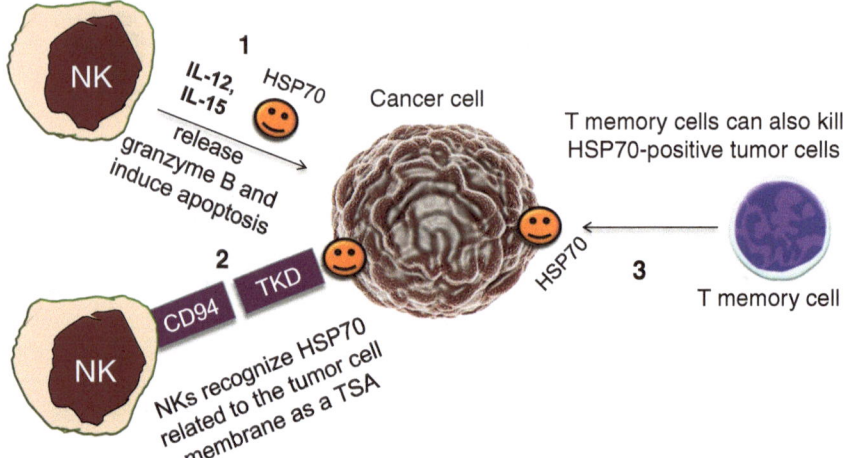

Fig. 6.6 Extracellular HSP70 can influence the immune system even in the absence of an antigenic peptide

antigen-presenting cells. However, currently nobody knows why natural killers attack a cell if HSP70 is integrated into the membrane, and do not attack the cell if HSP70 is bound to the cell through a receptor even though the TKD sequence of the receptor-bound HSP70 is readily available for interaction.

Although many questions remain to be answered, it is clear that HSP70 has a dual role that is dependent on its intra- and extracellular localization. Tumour cells use intracellular HSP70, and this HSP70 is not recognized protects the cancer cells. In contrast, extracellular HSP70, is something more like a "prisoner" who has escaped from the "cancer jail" to inform the immune system to act. Here, natural killer cells and T cells recognize cancerous cells from normal healthy cells, and therefore act to destroy these cells.

6.3 Unresolved Issues of Carcinogenesis, or How to Make a Million Dollars (and P.S.)

A significant amount of knowledge is known about the mechanisms of carcinogenesis and the role HSP70 plays within these mechanisms. However, cancer still exists and modern medicine struggles to find suitable solutions to treat such diseases. This indicates that there are important aspects of tumour cells that we do not understand! This is not just about how smart our reasoning is, or how "generously" the state funds science; the inherent complexity of tumour cells still exceeds the capabilities of our knowledge on this topic. This sobering conclusion; however, should not overshadow the fact that we have made enormous advances in the study of tumours, and we can expect a rapid, if not sensational breakthrough in the next decade!

However, studies of tumours are currently undergoing a substantial reassessment. Understanding the role of stem tumour cells, precursor cells and differentiated tumour cells can change our view of carcinogenesis (Reya and Clevers 2005; Reya et al. 2001). The role of HSP70 in stem cell growth of cancer cells remains completely unknown; however, understanding the role of HSP70 in these cells may become highly significant for any future approaches aimed at treating cancer.

P.S. At this point, it would be appropriate to recall a historical fact. The will of Alfred Nobel, the founder of the Nobel Foundation, says that when the problem of carcinogenesis is solved, all the money in the fund (and that is a lot of money; many millions of dollars!) should be given to the person who solved the problem of cancer. Now, I have an important statement to make! If ever any of you dear readers solve the problem of carcinogenesis, remember who told you about tumours in this chapter!

References

Afanasyev (1873) Narodnye russkie skazki (Russian fairy tale)

Afanasyeva EA, Komarova EY, Larsson LG, Bahram F, Margulis BA, Guzhova IV (2007) Drug-induced Myc-mediated apoptosis of cancer cells is inhibited by stress protein Hsp70. Int J Cancer 121(12):2615–2621

Ansieau S, Bastid J, Doreau A (2008) Induction of EMT by twist proteins as a collateral. Effect of tumor-promoting inactivation of premature senescence. Cancer Cell 14:79–89

Barnes JA, Dix DJ, Collins BW, Luft C, Allen JW (2001) Expression of inducible Hsp70 enhances the proliferation of MCF-7 breast cancer cells and protects against the cytotoxic effects of hyperthermia. Cell Stress Chaper 6:316–325

Beere HM, Wolf BB, Cain K et al (2000) Heat-shock protein 70 inhibits apoptosis by preventing recruitment of procaspase-9 to the Apaf-1 apoptosome. Nat Cell Biol 2(8):469–475

Benanti JA, Galloway DA (2004) The normal. Response to RAS: senescence or transformation? Cell Cycle 3:715–717

Bennett M, Macdonald K, Chan SW, Luzio JP, Simari R, Weissberg P (1998) Cell surface trafficking of Fas: a rapid mechanism of p53-mediated apoptosis. Science 282:290–293

Bihani T, Mason DX, Jackson TJ, Chen SC, Boettner B, Lin AW (2004) Differential oncogenic Ras signaling and senescence in tumor cells. Cell Cycle 3:1201–1207

Blyth K, Stewart M, Bell M, James C, Evan G, Neil JC, Cameron ER (2000) Sensitivity to myc-induced apoptosis is retained in spontaneous and transplanted lymphomas of CD2-mycER (TM) mice. Oncogene 19:773–782

Bouvard V, Zaitchouk T, Vacher M et al (2000) Tissue and cell-specific expression of the p53-target genes: bax, fas, mdm2 and waf1/p21, before and following ionising irradiation in mice. Oncogene 19:649–660

Braig M, Schmitt CA (2006) Oncogene-induced senescence: putting the brakes on tumor development. Cancer Res 66:2881–2884

Calderwood SK, Khaleque MA, Sawyer DB, Ciocca DR (2006) Heat shock proteins in cancer: chaperones of tumorigenesis. Trends Biochem Sci 31:164–172

Campisi J (2005) Aging, tumor suppression and cancer: high wire-act! Mech Aging Dev 126:51–58

Chen Z, Trotman LC, Shaffer D et al (2005) Crucial role of p53-dependent cellular senescence in suppression of Pten-deficient tumorigenesis. Nature 436:725–730

Ciocca DR, Calderwood SK (2005) Heat shock proteins incancer:diagnostic, prognostic, predictive, and treatment implications. Cell Stress Chaperon 10:86–103

Ciocca DR, Clark GM, Tandon AK, Fuqua SA, Welch WJ, McGuire WL (1993) Heat shock protein hsp70 in patients with axillary lymph node-negative breast cancer: prognostic implications. J Natl Cancer Inst 85:570–574

Costa M, Rosas S, Chindano A, Lima P, Madi K, Carvalho M (1997) Expression of heat shock protein 70 and P53 in human lung cancer. Oncol Rep 4:1113–1116

Creagh EM, Carmody RJ, Cotter TG (2000) Heat shock protein 70 inhibits caspase-dependent and -independent apoptosis in Jurkat T cells. Exp Cell Res 257:58–66

Di Micco R, Fumagalli M, Cicalese A et al (2006) Oncogene-induced senescence is a DNA damage response triggered by DNA hyperreplication. Nature 444:638–642

Diehl JA, Yang W, Rimerman RA, Xiao H, Emili A (2003) Hsc70 regulates accumulation of cyclin D1 and cyclin D1-dependent protein kinase. Mol Cell Biol 23:1764–1774

Dix DJ, Allen JW, Collins BW et al (1996) Targeted gene disruption of Hsp70-2 results in failed meiosis, germ cell apoptosis, and male infertility. Proc Natl Acad Sci USA 93:3264–3268

Ehrhart JC, Duthu A, Ullrich S, Appella E, May P (1988) Specific interaction between a subset of the p53 protein family and heat shock proteins hsp72/hsc73 in a human osteosarcoma cell line. Oncogene 3:595–603

Eischen CM, Weber JD, Roussel MF, Sherr CJ, Cleveland JL (1999) Disruption of the ARF-Mdm2-p53 tumor suppressor pathway in Myc-induced lymphomagenesis. Genes Dev 13:2658–2669

Ferbeyre G, de Stanchina E, Lin AW et al (2002) Oncogenic ras and p53 cooperate to induce cellular senescence. Mol Cell Biol 22:3497–3508

Gabai VL, Meriin AB, Yaglom JA, Wei JY, Mosser DD, Sherman MY (2000) Suppression of stress kinase JNK is involved in HSP72- mediated protection of myogenic cells from transient energy deprivation. HSP72 alleviates the stress-induced inhibition of JNK dephosphorylation. J Biol Chem 275:38088–38094

Garbe JC, Holst CR, Bassett E, Tlsty T, Stampfer MR (2007) Inactivation of p53 function in cultured human mammary epithelial cells turns the telomere-length dependent senescence barrier from agonescence into crisis. Cell Cycle 6:1927–1936

Garrido C, Fromentin A, Bonnotte B et al (1998) Heat shock protein 27 enhances the tumorigenicity of immunogenic rat colon carcinoma cell clones. Cancer Res 58:5495–5499

Gehrmann M, Schmetzer H, Eissner G et al (2003) Membrane-bound heat shock protein 70 (Hsp70) in acute myeloid leukemia: a tumor specific recognition structure for the cytolytic activity of autologous NK cells. Haematologica 88(4):474–476

Gehrmann M, Liebisch G, Schmitz G et al (2008) Tumor-specific Hsp70 plasma membrane localization is enabled by the glycosphingolipid Gb3. PLoS ONE 3(4):e1925

Green DR, Kroemer G (2009) Cytoplasmic functions of the tumour suppressor p53. Nature 458(7242):1127–1130

Gross C, Koelch W, DeMaio A, Arispe N, Multhoff G (2003) Cell surface-bound heat shock protein 70 (Hsp70) mediates perforin-independent apoptosis by specific binding and uptake of granzyme B. J Biol Chem 278(42):41173–41181

Guo F, Sigua C, Bali P et al (2005) Mechanistic role of heat shock protein 70 in Bcr-Abl-mediated resistance to apoptosis in human acute leukemia cells. Blood 105:1246–1255

Gurbuxani S, Schmitt E, Cande C et al (2003) Heat shock protein 70 binding inhibits the nuclear import of apoptosis-inducing factor. Oncogene 22:6669–6678

Hahn WC, Weinberg RA (2002) Rules for making human tumor cells. N Engl J Med 347(20):1593–1603

Hanahan D, Weinberg RA (2000) The hallmarks of cancer. Cell 100:57–70

Hayflick L (1979) Cell biology of aging. Fed Proc 38:1847–1850

Horváth I, Vígh L (2010) Cell biology: stability in times of stress. Nature 463(7280):436–438

Horváth I, Multhoff G, Sonnleitner A, Vígh L (2008) Membrane-associated stress proteins: more than simply chaperones. Biochim Biophys Acta 1778(7–8):1653–1664

Jäättelä M (1995) Over-expression of hsp70 confers tumorigenicity to mouse fibrosarcoma cells. Int J Cancer 60:689–693

Jäättelä M (1999) Escaping cell death: survival proteins in cancer. Exp Cell Res 248:30–43

Jäättelä M, Wissing D, Kokholm K, Kallunki T, Egeblad M (1998) Hsp70 exertsitsanti-apoptotic function downstream of caspase-3-like proteases. EMBO J 17:6124–6134

Jiang B, Xiao W, Shi Y, Liu M, Xiao X (2005) Heat shock pretreatment inhibited the release of Smac/DIABLO from mitochondria and apoptosis induced by hydrogen peroxide in cardio-myocytes and C2C12 myogenic cells. Cell Stress Chaperon 10(3):252–262

Kalinowska M, Garncarz W, Pietrowska M, Garrard WT, Widlak P (2005) Regulation of the human apoptotic DNase/RNase Endonuclease G: involvement of Hsp70 and ATP. Apoptosis 10:821–830

Kelly S, Zhang ZJ, Zhao H et al (2002) Gene transfer of HSP72 protects cornu ammonis 1 region of the hippocampus neurons from global ischemia: influence of Bcl-2. Ann Neurol 52(2):160–167

Kim NW, Piatyszek MA, Prowse KR et al (1994) Specific association of human telomerase activity with immortal cells and cancer. Science 266:2011–2015

Laad AD, Thomas ML, Fakih AR, Chiplunkar SV (1999) Human gamma delta T cells recognize heat shock protein-60 on oral tumor cells. Int J Cancer 80(5):709–714

Lee SH, Kwon HM, Kim YJ, Lee KM, Kim M, Yoon BW (2004) Effects of hsp70.1 gene knockout on the mitochondrial apoptotic pathway after focal cerebral ischemia. Stroke 35(9):2195–2199

Lee JS, Lee JJ, Seo JS (2005) HSP70 deficiency results in activation of c-Jun N-terminal. Kinase, extracellular signal-regulated kinase, and caspase-3 in hyperosmolarity-induced apoptosis. J Biol Chem 280(8):6634–6641

Lee SH, Lee SJ, Jung YS, Xu Y, Kang HS, Ha NC, Park BJ (2009) Blocking of p53-Snail binding, promoted by oncogenic K-Ras, recovers p53 expression and function. Neoplasia 11:22–31

Leu JI, Dumont P, Hafey M, Murphy ME, George DL (2004) Mitochondrial p53 activates Bak and causes disruption of a Bak-Mcl1 complex. Nat Cell Biol 6(5):443–450

Li CY, Lee JS, Ko YG, Kim JI, Seo JS (2000) Heat shock protein 70 inhibits apoptosis downstream of cytochrome c release and upstream of caspase-3 activation. J Biol Chem 275:25665–25671

Li Y, Pan J, Li JL et al (2007) Transcriptional changes associated with breast cancer occur as normal human mammary epithelial cells overcome senescence barriers and become immortalized. Mol Cancer 6:7

Mason DX, Jackson TJ, Lin AW (2004) Molecular signature of oncogenic ras-induced senescence. Oncogene 23(57):9238–9246

Matsuda Y (2008) Molecular mechanism underlying the functional loss of cyclindependent kinase inhibitors p16 and p27 in hepatocellular carcinoma. World J Gastroenterol 14:1734–1740

Matsumori Y, Hong SM, Aoyama K (2005) Hsp70 overexpression sequesters AIF and reduces neonatal hypoxic/ischemic brain injury. J Cereb Blood Flow Metab 25:899–910

Matsumori Y, Northington FJ, Hong SM (2006) Reduction of caspase-8 and -9 cleavage is associated with increased c-FLIP and increased binding of Apaf-1 and Hsp70 after neonatal hypoxic/ischemic injury in mice overexpressing Hsp70. Stroke 37(2):507–512

McCormick JJ, Maher VM (1988) Towards an understanding of the malignant transformation of diploid human fibroblasts. Mutat Res 199:273–291

Mihara M, Erster S, Zaika A et al (2003) p53 has a direct apoptogenic role at the mitochondria. Mol Cell 11(3):577–590

Moser C, Schmidbauer C, Gürtler U et al (2002) Inhibition of tumor growth in mice with severe combined immunodeficiency is mediated by heat shock protein 70 (Hsp70)-peptide-activated, CD94 positive natural killer cells. Cell Stress Chaperon 7(4):365–373

Müller M, Wilder S, Bannasch D et al (1998) p53 activates the CD95 (APO-1/Fas) gene in response to DNA damage by anticancer drugs. J Exp Med 188:2033–2045

Multhoff G (2002) Activation of natural killer cells by heat shock protein 70. Int J Hyperthermia 6:576–585

Multhoff G (2007) Heat shock protein 70 (Hsp70): membrane location, export and immunological relevance. Methods 43(3):229–237

Multhoff G, Botzler C, Wiesnet M et al (1995) A stress-inducible 72 kDa heat shock protein (Hsp72) is expressed on the surface of human tumor cells, but not on normal cells. Int J Cancer 61:272–279

Multhoff G, Botzler C, Jennen L, Schmidt J, Ellwart J, Issels R (1997) Heat shock protein 72 on tumor cells: a recognition structure for natural killer cells. J Immunol 158(9):4341–4350

Multhoff G, Mizzen L, Winchester CC et al (1999) Heat shock protein 70 (Hsp70) stimulates proliferation and cytolytic activity of natural killer cells. Exp Hematol 27(11):1627–1636

Multhoff G, Pfister K, Gehrmann M et al (2001) A 14-mer Hsp70 peptide stimulates natural killer (NK) cell activity. Cell Stress Chaperon 6(4):337–344

Nanbu K, Konishi I, Mandai M, Kuroda H, Hamid AA, Komatsu T, Mori T (1998) Prognostic significance of heat shock proteins Hsp70 and Hsp90 in endometrial carcinomas. Cancer Detect Prev 22:549–555

Nihei T, Sato N, Takahashi S et al (1993) Demonstration of selective protein complexes of p53 with 73 kDa heat shock cognate protein, but not with 72 kDa heat shock protein in human tumor cells. Cancer Lett 73:181–189

Nilsson JA, Cleveland JL (2003) Myc pathways provoking cell suicide and cancer. Oncogene 22(56):9007–9021

Nylandsted J, Brand K, Jäättelä M (2000a) Heat shock protein 70 is required for the survival of cancer cells. Ann NY Acad Sci 926:122–125

Nylandsted J, Rohde M, Brand K, Bastholm L, Elling F, Jäättelä M (2000b) Selective depletion of heat shock protein 70 (Hsp70) activates a tumor-specific death program that is independent of caspases and bypasses Bcl-2. Proc Natl Acad Sci USA 97:7871–7876

Olsen CL, Gardie B, Yaswen P, Stampfer MR (2002) Raf-1-induced growth arrest in human mammary epithelial cells is p16-independent and is overcome in immortal cells during conversion. Oncogene 21:6328–6339

Park HS, Lee JS, Huh SH, Seo JS, Choi EJ (2001) Hsp72 functions as a natural inhibitory protein of c-Jun N-terminal kinase. EMBO J 20(3):446–456

Peeper DS, Dannenberg JH, Douma S, te Riele H, Bernards R (2001) Escape from premature senescence is not sufficient for oncogenic transformation by Ras. Nat Cell Biol 3(2):198–203

Pelengaris S, Rudolph B, Littlewood T (2000) Action of Myc in vivo-proliferation and apoptosis. Opin Gen Dev 10:100–105

Perez-Sala D, Rebollo A (1999) Novel aspects of Ras proteins biology: regulation and implications. Cell Death Diff 6:722–728

Pfister K, Radons J, Busch R et al (2007) Patient survival by Hsp70 membrane phenotype: association with different routes of metastasis. Cancer 110(4):926–935

Prendergast GC (1999) Mechanism of apoptosis by Myc. Oncogene 18:2967–2987

Ravagnan L, Gurbuxani S, Susin SA et al (2001) Heat-shock protein 70 antagonizes apoptosis-inducing factor. Nat Cell Biol 3(9):839–843

Reya T, Clevers H (2005) Wnt signalling in stem cells and cancer. Nature 434(7035):843–850

Reya T, Morrison SJ, Clarke MF, Weissman IL (2001) Stem cells, cancer, and cancer stem cells. Nature 414(6859):105–111

Romanucci M, Bastow T, Della Salda L (2008) Heat shock proteins in animal neoplasms and human tumours–a comparison. Cell Stress Chaperon 13(3):253–262

Ruchalski K, Mao H, Li Z et al (2006) Distinct hsp70 domains mediate apoptosis-inducing factor release and nuclear accumulation. J Biol Chem 281:7873–7880

Saleh A, Srinivasula SM, Balkir L, Robbins PD, Alnemri ES (2000) Negative regulation of the Apaf-1 apoptosome by Hsp70. Nat Cell Biol 2(8):476–483

Santarosa M, Favaro D, Quaia M, Galligioni E (1997) Expression of heat shockprote in 72 in renal cell carcinoma: possible role and prognostic implications in cancer patients. Eur J Cancer 33:873–877

Schmitt E, Gehrmann M, Brunet M, Multhoff G, Garrido C (2007) Intracellular and extracellular functions of heat shock proteins: repercussion in cancer therapy. Leuco Biol 81:15–27

Schwabe M, Lubbert M (2007) Epigenetic lesions in malignant melanoma. Curr Pharm Biotechnol 8:382–387

Sebastian T, Johnson PF (2006) Stop and go: antiproliferative and mitogenic functions of the transcription factor C/EBP. Cell Cycle 5:953–957

Seo JS, Park YM, Kim JI et al (1996) T cell lymphoma in transgenic mice expressing the human Hsp70 gene. Biochem Biophys Res Comm 218:582–587

Shay JW, Bacchetti S (1997) A survey of telomerase activity in human cancer. Eur J Cancer 33:787–791

Sherman M (2010) Major heat shock protein Hsp72 controls oncogene-induced senescence. Ann N Y Acad Sci 1197:152–157

Shin BK, Wang H, Yim AM et al (2003) Global profiling of the cell surface proteome of cancer cells uncovers an abundance of proteins with chaperone function. J Biol Chem 278:7607–7616

Srivastava P (2002) Interaction of heat shock proteins with peptides and antigen presenting cells: chaperoning of the innate and adaptive immune responses. Annu Rev Immunol 20:395–425

Srivastava PK, Menoret A, Basu S, Binder RJ, McQuade KL (1998) Heat shock proteins come of age: primitive functions acquire new roles in an adaptive world. Immunity 8(6):657–665

Stankiewicz AR, Lachapelle G, Foo CP, Radicioni SM, Mosser DD (2005) Hsp70 inhibits heat-induced apoptosis upstream of mitochondria by preventing Bax translocation. J Biol Chem 280(46):38729–38739

Steel R, Doherty JP, Buzzard K et al (2004) Hsp72 inhibits apoptosis upstream of the mitochondria and not through interactions with Apaf-1. J Biol Chem 279(49):51490–51499

Sun Y, Ouyang YB, Xu L et al (2006) The carboxyl-terminal. domain of inducible Hsp70 protects from ischemic injury in vivo and in vitro. J Cereb Blood Flow Metab 26(7):937–950

Tang D, Khaleque MA, Jones EL et al (2005) Expression of heat shock proteins and heat shock protein messenger ribonucleic acid in human prostate carcinoma in vitro and in tumors in vivo. Cell Stress Chaperon 10(1):46–58

Triantafilou M, Triantafilou K (2004) HSP70 and HSP90 associate with Toll-like receptor 4 in response to bacterial lipopolysaccharide. Biochem Soc Trans 32(Pt 4):636–639

Trieb K, Lechleitner T, Lang S, Windhager R, Kotz R, Dirnhofer S (1998) Heat-shock protein 72 expression in osteosarcomas correlates with good response to neoadjuvant chemotherapy. Hum Pathol 29:1050–1055

Trost TM, Lausch EU, Fees SA et al (2005) Premature senescence is a primary fail-safe mechanism of ERBB2-driven tumorigenesis in breast carcinoma cells. Cancer Res 65:840–849

Tsuchiya D, Hong S, Matsumori Y (2003) Overexpression of rat heat shock protein 70 is associated with reduction of early mitochondrial cytochrome C release and subsequent DNA fragmentation after permanent focal ischemia. J Cereb Blood Flow Metab 23(6):718–727

Vargas-Roig LM, Fanelli MA et al (1997) Heat shock proteins and cell proliferation in human breast cancer biopsy samples. Cancer Detect Prev 21:441–451

Vaseva AV, Moll UM (2009) The mitochondrial p53 pathway. Biochim Biophys Acta 1787(5):414–420

Vega VL, Rodríguez-Silva M, Frey T et al (2008) Hsp70 translocates into the plasma membrane after stress and is released into the extracellular environment in a membrane-associated form that activates macrophages. J Immunol 180(6):4299–4307

Voellmy R (2004) On mechanisms that control heat shock transcription factor activity in metazoan cells. Cell Stress Chaperon 9(2):122–133

Vogelstein B, Kinzler KW (1993) The multistep nature of cancer. Trends Genet 9:138–141

Volloch VZ, Sherman (1999) Oncogenic potential of Hsp72. Oncogene 18:3648–3651

Wei YQ, Zhao X, Kariya Y, Teshigawara K, Uchida A (1995) Inhibition of proliferation and induction of apoptosis by abrogation of heat-shock protein (Hsp) 70 expression in tumor cells. Cancer Immunol Immunother 40:73–77

White E (1998) Regulation of apoptosis by adenovirus E1A and E1B oncogenes. Semin Virol 8:505–513

Wright WE, Shay JW (2001) Cellular senescence as a tumor-protection mechanism: the essential role of counting. Curr Opin Genet Dev 11:98–103

Wu GS, Burns TF, McDonald ER 3rd et al (1997) KILLER/DR5 is a DNA damage-inducible p53-regulated death receptor gene. Nat Genet 17:141–143

Yaswen P, Campisi J (2007) Oncogene-induced senescence pathways weave an intricate tapestry. Cell 128:233–234

Yu K, Ravera CP, Chen YN, McMahon G (1997) Regulation of Myc-dependent apoptosis by P53, C-Jun N-terminal. kinases stress-activated protein kinases, and Mdm-2. Cell Growth Diff 8:731–742

Zeng Y, Chen X, Larmonier N et al (2006) Natural killer cells play a key role in the anti-tumor immunity generated by chaperone-rich cell lysate vaccination. Int J Cancer 119(11):2624–2631

Zhu J, Woods D, McMahon M, Bishop JM (1998) Senescence of human fibroblasts induced by oncogenic Raf. Genes Dev 12(19):2997–3007

Chapter 7
HSP70 in Aging

Abstract Aging or senescence in biology is defined as the process of gradual loss of important body functions and cells, and in particular, the inability of cells to reproduce and cope with stress. Important features of an aging cell include the progressive accumulation of damaged proteins, abnormal protein aggregates and oxidative stress. Such features may damage macromolecules and trigger two genetic programs—cellular senescence and apoptosis. HSP70 limits cellular aging by: (i) ensuring the refolding and disaggregation of denatured/misfolded proteins; (ii) participating in the degradation of irreversibly dysfunctional proteins; (iii) mediating the effects of histone deacetylase 6 in the starvation-induced increase in life expectancy; and (iv) preventing cell senescence and apoptosis. However, cells of an aging body display a dramatic reduction of HSP70 inducibility. This reduction correlates with a decrease in the ability of cells to cope with stress. The reduction of HSP70 inducibility most likely reflects the launch of a special genetic program aimed at the activation of JNK-dependent apoptosis and the destruction of old cells, which have accumulated damaged proteins and dangerous mutations. Such a program functions to protect the body as a whole.

Keywords HSP70 • Age • Oxidative stress • Histone deacetylase 6 • JNK kinase

The interest of people in the topic of this chapter is significant. Frankly speaking, for many, it is even bigger than the interest in the origin of life! Why do we age so quickly and why do we have such short lives? In this chapter, I will talk about the mechanisms of aging and longevity, and more specifically, about the role of HSP70 in these mechanisms.

When scientists studied the role of HSP70 in folding, refolding, and the maintenance of protein homeostasis, it was of significant interest for a number of molecular biologists. For most people, this research was unfamiliar and did not seem relevant to medicine. However, as it became apparent that HSP70 increases the cells' resistance to various injuries can block cell aging and even cell death, researchers began to consider that these proteins may play a significant role in the mechanisms of human longevity. These intuitive assumptions proved to be correct. Heat shock proteins begun to be of interest not only to scientists, but to all people who wished for a longer, healthier life, rather than becoming a burden for their own children in old age!

I. Malyshev, *Immunity, Tumors and Aging: The Role of HSP70*,
SpringerBriefs in Biochemistry and Molecular Biology,
DOI: 10.1007/978-94-007-5943-5_7, © The Author(s) 2013

7.1 What is Senescence and its Underlying Causes

Aging or senescence in biology refers to the process of gradual disruption and loss of important body functions. This process also refers to cells losing the ability to reproduce and regenerate, and to cope with stress and damage factors. Due to senescence, the body and individual cells become less adapted to environmental conditions, and thus, their resistance to stress and the action of damaging factors decreases. The rate of aging of an individual cell and the body generally reflects a conflict between two forces: the effects of different stressors and harmful factors that attack the cell throughout its life, and body defence mechanisms that maintain homeostasis, vitality and longevity.

Two main factors play a most important role in the aging of a cell: (1) progressive accumulation of damaged proteins and abnormal toxic protein aggregates, and (2) oxidative stress, which may lead to damage of macromolecules and trigger two genetic programs—the program of cellular senescence and the program of cell death, apoptosis.

In all of these cases, HSP70 restricts the development of the aging processes. However, the aging process does result in a decrease in the activity of HSP70 induction systems. Therefore the key question of the concept of aging is: why do HSP70 induction systems decrease in an aging cell?

7.2 Accumulation of Damaged Proteins, HSP70 and the Sacrificial Altruism of the Ageing Daughter Cells

Starting at mitotic birth and throughout its life, the cell is constantly exposed to various unfavourable stress factors. The most common impact of these factors is damage and denaturation of cellular proteins and, consequently, formation of toxic protein aggregates. Progressive accumulation of damaged proteins and aggregates plays a key role in disrupting cellular function, and in accelerating the aging of cells and the body as a whole.

Fortunately, these adverse age-related changes can be effectively limited for a long time by the system of protein quality control and maintenance of protein homeostasis (Garrido et al. 2006; Lindquist and Craig 1988), i.e., by the special **FORD** mechanism. It would be useful to remind ourselves how cells respond to the disruption of protein homeostasis (see Chap. 3) and focus our discussion on those aspects that are related to aging.

A damaged protein appears in a young or senescent cell. What happens to it next? The choice is limited to four pathways, but a disruption to any of these pathways will accelerate the process of aging.

First, the protein quality control system will attempt to send the protein to refolding and restore its native structure (Figs. 1.6 and 2.6). HSP70, its HSP40 co-chaperones and the factors of nucleotide exchange, Bag-1 and HSPBP-1, play a key role in this process (Min et al. 2008).

Second, the quality control system will send the irreversibly damaged proteins to degradation via proteasomes (Fig. 2.4 and 2.6). HSP70, HSP40, CHIP ubiquitin ligase and proteasomes play an important role in proteosomal cleavage of the protein (Marques et al. 2006). In this case, the denatured protein, scavenged by the HSP70 protein, moves closer to CHIP, and becomes polyubiquitinated with its help, thus becoming a target for degradation in a proteasome.

Third, in lysosomal degradation, HSP70 recognizes a label block, KFERQ, on the protein intended for degradation (Fig. 2.5), binds to this protein, and delivers the protein into the lysosome lumen for degradation (Dice 2007).

These three mechanisms maintain protein homeostasis when denatured proteins appear in a cell. If something happens, and, for whatever reasons, these mechanisms fail, the denatured proteins begin associating to form toxic aggregates. In cells this leads to disruption of cellular functions and death (Winklhofer et al. 2008; Hands et al. 2008).

However, even in this case, HSP70 can save the situation by a fourth mechanism. Together with the HSP100 protein, HSP70 can pull protein aggregates apart (Hut et al. 2005; Mayer and Bukau 2005). As a result, each denatured protein released from the aggregate gets a second chance of refolding or is sent for degradation (Fig. 3.2).

Molecular chaperones and their helpers are especially important during aging, when the number of damaged proteins in the cell increases. For example, inactivation of CHIP in mice leads to a significant reduction in life expectancy (Min et zal. 2008), and a decrease in HSP70-dependent lysosomal degradation is associated with the development of Parkinson's disease (Dice 2007).

Thus, HSP70 activity in refolding, disaggregation and degradation of irreparable proteins forms the basis for the anti-aging effects of the protein quality control system.

However, the ability of cells to refold or degrade misfolded/dysfunctional proteins reduces as we grow older, while the number of damaged denatured proteins in the cell increases. Consequently, a point is reached when the protein quality control system, the FORD machinery, can no longer cope with the level of accumulated damaged proteins and protein aggregates. At this point, interesting, almost dramatic events begin.

When the FORD machinery begins to be seriously disrupted, protein aggregates begin being actively transported along microtubules to centrosomes and accumulate there in the form of aggresomes (Rujano et al. 2006). Aggresomes are then removed either by macroautophagy or, more often, in asymmetrical division (Fig. 7.1). The result of asymmetrical division is that one daughter cell is cleared of all aggregates, whereas the other inherits aggresomes! The first cell then becomes "healthy" and continues functioning normally, while the second, overloaded by aggresomes, is destroyed by apoptosis, basically "sacrificed for the health" of its sister cell! (Rujano et al. 2006).

In mammals, the formation of aggresomes is regulated by histone deacetylase 6 (HDAC6) (Fig. 7.2). Histone deacetylases similar to HDAC6 are very interesting enzymes. Histone deacetylases catalyse cleavage of acetyl groups from

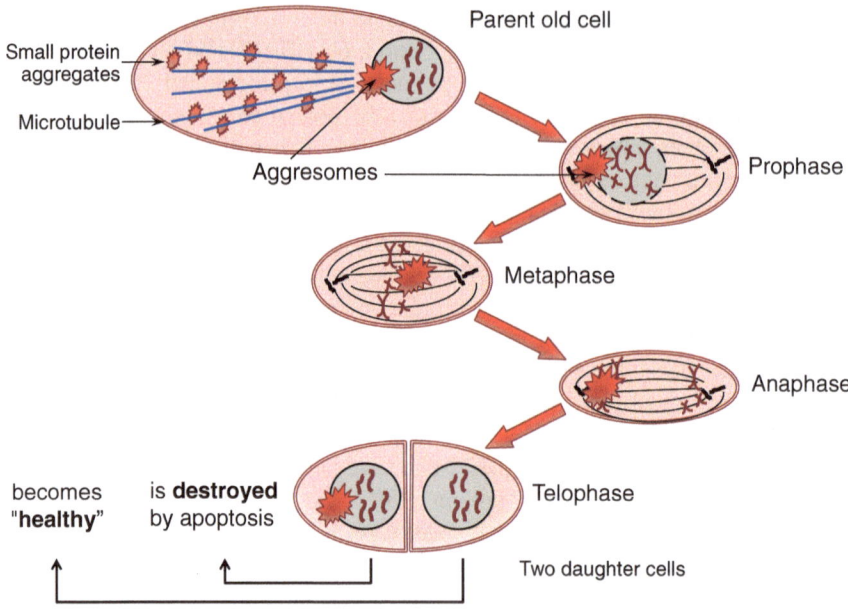

Fig. 7.1 Aggresome accumulation and their removal in asymmetrical cell division

Fig. 7.2 Histone deacetylase 6 (HDAC6) regulates the formation of aggresomes and HSF-1 activation and due to this effect HDAC6 is involved in the beneficial effects of a reduced calorie diet

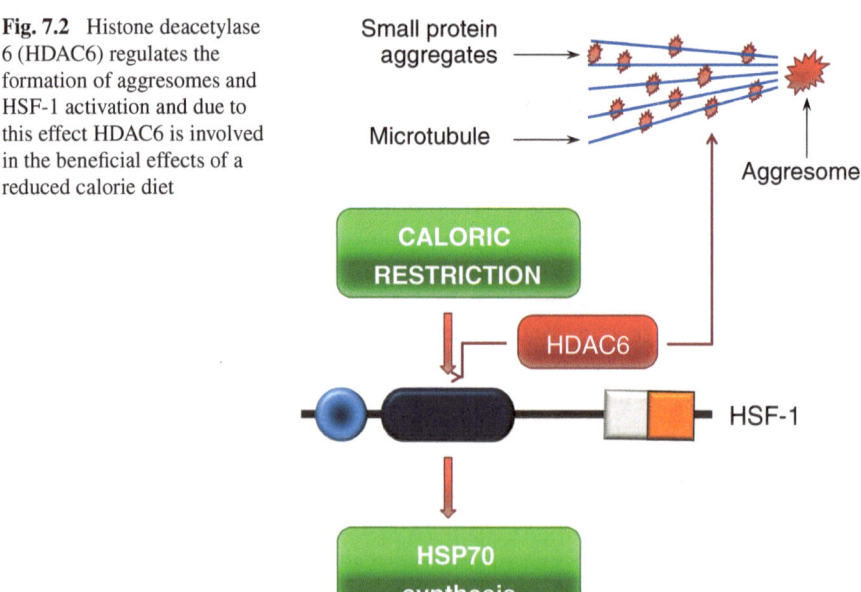

lysine residues of many substrates. HDAC6 can bind both damaged proteins and microtubules (Boyault et al. 2007), thus providing the transfer of protein aggregates along microtubules to centrosomes. Additionally, owing to deacetylation of histones in the protein microenvironment of the *HSP70* gene, this enzyme can induce detachment of HSP90 from HSF1 (which sits on the promoter) thus activating HSF1 and HSP70 synthesis (Boyault et al. 2007; Kovacs et al. 2005; Rujano et al. 2006; Westerheide et al. 2009).

It is well known that a reduced calorie diet may prolong our life and improve our health (Corbi et al. 2012; Sherman et al. 2011; Dall et al. 2009; Everitt et al. 2007). Genetic studies have shown that HDACs play a key role in this interesting effect (Westphal et al. 2007). The starvation-associated increased activity of HSF1 (Steinkraus et al. 2008) and increased life expectancy of *C. elegans* are blocked when HDACs are inhibited. The induction of HDACs by starvation was shown to increase the life expectancy of the worm *Caenorhabditis elegans* and the fruit fly *Drosophila melanogaster* (Westphal et al. 2007). This effect of HDACs is associated with the separation of aggresomes during asymmetrical division and activation of HSP70 synthesis.

7.3 Oxidative Stress and HSP70: The Unity and the Struggle of Opposites

In addition to the accumulation of damaged, denatured proteins, increases in the levels of reactive oxygen species (Powis et al. 1995) leads to increases in oxidative damage to DNA, lipids and proteins. This oxidative damage plays an important role in cell damage and accelerated ageing. HSP70 may limit the oxidative stress-induced damage to proteins through mechanisms that we have discussed: due to refolding, degradation and disaggregation of damaged proteins.

At the same time, the role of oxidative stress in aging is not limited to damaged intracellular macromolecules. It is more specific (Fig. 7.3). Oxidative stress may accelerate aging and cell death through activation of apoptosis and the cell senescence program (Gabai et al. 1998). It was shown, for example, that in age-related neurodegenerative diseases such as Alzheimer's, it is precisely oxidative stress that triggers apoptosis, which is the main cause of neuronal death (Gabai et al. 1998).

Can the anti-aging effects of HSP70 manifest when cell aging and death of cells are triggered by signalling mechanisms, rather than by the accumulation of denatured proteins? This question was experimentally tested and HSP70 was found to prevent apoptosis induced by factors such as radiation (Simon et al. 1995; Raffray and Cohen 1997; Cosulich and Clarke 1996; Mosser and Martin 1992), TNF-α signalling molecules (Jäättelä 1993) and NO (Bellmann et al. 1996).

The material presented in the previous chapter will help us understand how HSP70 does this. I talked about how the activation of both apoptosis and senescence goes through two consecutive key steps: the activation of the JNK stress-kinase and the activation of the p53 transcription factor. We already know from the previous chapter that HSP70 can block both steps, as well as some other steps in apoptosis (Cosulich and Clarke 1996; Gabai et al. 1997; Gabai et al. 1998; Mignotte and Vayssiere 1998; Salvesen and Dixit 1997; Seimiya et al. 1997;

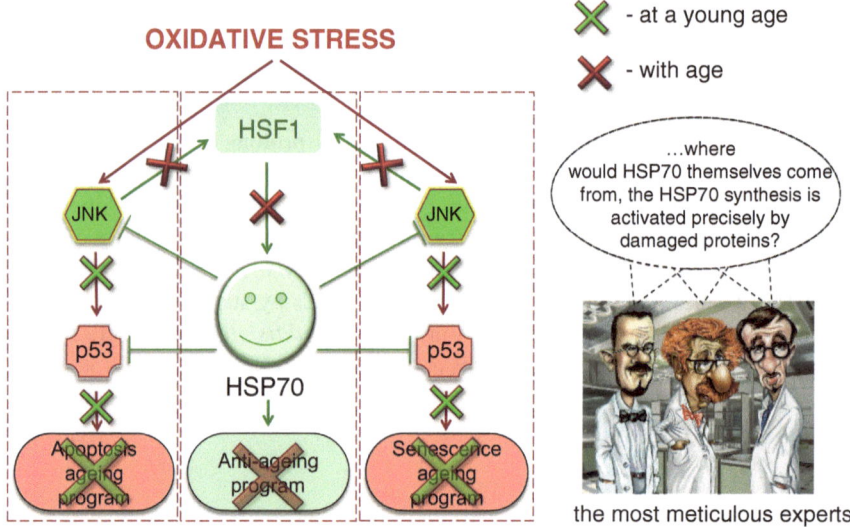

Fig. 7.3 Oxidative stress can, through activation of JNK kinase, simultaneously trigger two opposing programs in the same cell: the aging program and the HSP70-dependent anti-aging program

Xia et al. 1995; Mosser et al. 1997; Verheij et al. 1996; Volloch et al. 1998; Zanke et al. 1996; Webb et al. 1997). It is important to note that the suppression of JNK activation does not require participation of the ATP-ase HSP70 domain, i.e., it is ATP-independent. This means that HSP70 can perform this function in aging cells that may be experiencing a serious energy shortage.

Once it had been shown that HSP70 can block particular signalling pathways associated with aging and cell death in the absence of accumulated damaged proteins, the most meticulous experts were asking: "Excuse us, but where would HSP70 come from, when it is well known that HSP70 synthesis is activated precisely by damaged proteins?" It was necessary to find an answer to this question; otherwise the whole concept would be invalid. Dr. Mivechi's group had conducted studies and showed that in the absence of denatured protein, the stress-induced protein kinase JNK can itself activate the HSF1 transcription factor and thus the synthesis of HSP70 (Zanke et al. 1996; Simon et al. 1995).

Surprisingly, it appears that oxidative stress can, through activation of JNK kinase, simultaneously trigger two opposing programs in the same cell: the aging program and the HSP70-dependent anti-aging program. Apparently, a regulatory negative feedback mechanism forms immediately in the cell, and a certain balance between the JNK-dependent apoptosis and cellular aging programs on the one hand, and the activation of HSP70 synthesis on the other, is established.

The existence of such a balance between the pro-and anti-aging JNK programs sheds new light on another dark corner of aging. We can assume that the balance is shifted in favour of HSP70 at a young age. Therefore, despite strong oxidative stresses affecting a young cell, it can survive and divide, because the HSP70, activated by the same oxidative stress, blocks the program of JNK-dependent apoptosis and replicative aging. With age, when the inducibility of HSP70 is depressed, the balance is shifted in favour of the apoptosis and aging program. Therefore, even weak oxidative stress signals can activate mechanisms of replicative aging and apoptosis.

Now that we have examined the role HSP70 plays in aging cells, we can come to a justified conclusion: HSP70 is a key component of the intracellular system limiting cellular senescence. In this system HSP70:

- ensures refolding and disaggregation of denatured proteins that progressively accumulate in aging cells;
- participates in the degradation of irreversibly damaged proteins;
- mediates the effects of HDACs in the starvation-induced increase in life expectancy;
- prevents the development of the cell senescence program and apoptosis induced by oxidative stress.

7.4 Through HSP70 to Longevity, or "Eat Your Dinner Like a Pauper"

Increased levels of HSP70 have been correlated with increased life expectancy. Thus, HSP70 and HSF1 are very attractive pharmacological targets for anti-aging therapies. An increase in gene copies of HSP70 or HSF1 (Hsu et al. 2003), or starvation, which, through HDACs, stimulates both HSF1 activity and HSP70 synthesis, has been shown to increase the life span of *C. elegans* (Hsu et al. 2003) and *Drosophila* (Singh et al. 2007). You may be thinking that although such a system exists in flies, does a similar mechanism exist in humans? Do not worry; fortunately we are not evolutionarily that different from flies!

In humans, like in flies and worms, the potential for HSP70 induction noticeably declines with age (Singh et al. 2006). Aging, and age-related diseases and symptoms reflect precisely this reduction of HSP70 inducibility and resistance to stress. However, people who live over 100 years are an exception; their ability to induce HSP70 does not decrease with age (Ambra et al. 2004). If you want to know whether you will live to a 100 years, check your ability to induce HSP70 synthesis!

The more apparent it became that HSP70 played a critical role in limiting the aging process, the greater the number of pharmacologists that started raising the question of how to control HSP70 synthesis in the body in order to increase life expectancy was. A serious problem here is that the chemicals that induce the HSP70 synthesis, usually damage the cell, that is, they are cytotoxic, and cannot therefore be used by elderly people, whose cell resistance to stress factors is already reduced.

A glimmer of hope appeared when it was found that HDACs activated HSF1 in senescent cells. Consequently, using molecules that activate deacetylases could potentially activate expression of HSP70 genes. Such molecules are known: for example, resveratrol and dihydrocoumarin. These molecules are non-toxic in the doses that activate HDACs (Westphal et al. 2007; Westerheide et al. 2009). As such, pharmaceutical companies will conduct these studies and will spend significant sums of money on such drug development.

However, let us ignore the financial interests of pharmaceutical companies. Just recall that restricting calories and hunger may well activate HDACs, and thus activate the HSF1 transcription factor and HSP70 synthesis, and thus maintain homeostasis of proteins and block apoptosis in senescent cells. You can start devising ways of maintaining the inducibility of your chaperone system and increasing the duration of your life right now! In short, miss your dinner today, or eat it like a pauper! Or as they say in Russia, eat your breakfast yourself, share your lunch with your friend and give your supper to your enemy!

7.5 Why Does HSP70 Inducibility Decline with Age? A Chronicle of a Senescent Cell

Cells of an aging body display a dramatic reduction of HSP70 inducibility in response to stress (Heydari et al. 1994; Gabai and Kabakov 1993; Salvesen and Dixit 1997; Seimiya et al. 1997; Li et al. 1995; Njemini et al. 2002; Visala et al. 2003; Jin et al. 2004; Gutsmann-Conrad et al. 1998). Most often, inducibility is measured by an increase in the level of HSP70 in cells after heat shock at 42 °C (or 107.5 F). The decrease in this index closely correlates with a reduced ability to cope with environmental stress factors (Ames et al. 1993; Beckman and Ames 1998; Berlett and Stadtman 1997; Cortopassi and Wong 1999; Johnson et al. 1999; Kregel 2002; Kregel et al. 1995; Papaconstantinou 1994) and elevated rates of morbidity and mortality in older bodies subjected to periodic stress (Liu et al. 1996; Heydari et al. 1994; Hall et al. 2000; McArdle et al. 2004). On the contrary, the ability of the body to respond quickly to stress by HSP70 induction determines the high adaptive ability of the body and hence the ability to survive and maintain longevity (Minois et al. 2001).

There is a second question that is being asked more often: why does the ability to induce HSP70 synthesis in humans decrease with age? There is still no clear answer to this question. The age-related decline in the ability to activate HSP70 synthesis was found in nervous tissue (Sherman and Goldberg 2001; Winklhofer et al. 2008; Hands et al. 2008), skeletal and cardiac muscles (Kayani et al. 2008), and liver (Gagliano et al. 2007).

It is unclear why, but neurons of the central nervous system are much more prone to protein aggregation than cells in other organs and tissues. This could be due to significantly lower HSP70 inducibility in neurons when compared with other cells. *In neuronal tissue*, the age-related decline in activity of HSP70 synthesis in nerve cells is explained by the decline in expression of HSF1 and the reduced ability of HSF1 to form DNA-binding trimers (Batulan et al. 2003, Heydari et al. 1994). Inhibition of HSF1 activity and HSP70 synthesis occurs in Alzheimer's disease (Bhat et al. 2004) and it is believed that this is related to the accumulation of protein aggregates of β-amyloid peptide and the tau cytoskeletal protein in the brain of patients with Alzheimer's disease (Winklhofer et al. 2008).

The age-related decline in the ability to activate synthesis of heat shock proteins is also found in *muscle tissue*. In these tissues, strong contractions are a factor in the activation of HSP70 synthesis (Kayani et al. 2008). It is therefore assumed that the reduced contraction force in older animals and people may be related to a reduction of HSP70 inducibility (Kayani et al. 2008). As discussed before, starvation leads to the consecutive induction of HDACs and HSP70 synthesis, and thus contributes to increasing the life expectancy in worms and flies. We know now that exercise can induce synthesis of HSP70, the anti-aging protein. Thus, powerful biochemical and molecular biology studies have confirmed the slogan for those who want to live longer: "eat less and move more!"

The age-related decline in the HSP70 inducibility also occurs in liver cells (Gagliano et al. 2007). What are the consequences in this case? Heat shock proteins protect liver cells from toxic effects of alcohol, heavy metals, xenobiotics and oxidants (Lindquist and Craig 1988). Therefore, the age-related decline of HSP70 inducibility contributes significantly to reducing the detoxification function of the liver in older people (Gagliano et al. 2007).

Data obtained on yeast suggest that the age-related decline of HSP70 inducibility may be associated with a decrease in the activity of HDACs (Westphal et al. 2007), which activate HSF1 in senescent cells. It has also been shown that in an aging cell the CHIP protein begins making mistakes and sometimes starts ubiquitinating not only the damaged protein bound to HSP70, but also HSP70. As a result, HSP70 is also degraded in the proteasome. The observed increase in degradation of HSP70 may also contribute to the age-related decline in HSP70 inducibility (Min et al. 2008).

In thinking about the general causes of the age-dependent decline of HSP70 inducibility, it seems that these causes all represent special cases related to the effect of stress factors. Here is, however, one question that has not been answered: why in both body cells and in isolated cells placed in ideal conditions does HSP70 inducibility *inevitably* decrease? We do not know why it happens, but the very "*inevitability*" of the process clearly suggests that the age-related decline of HSP70 inducibility is genetically programmed. When a normal somatic cell stops dividing, we know that this is genetically programmed and is because of the reduced telomere length. Carol Greider recently won the Nobel Prize (2009) for the discovery of this mechanism. The genetic mechanism that determines the age-related decline in HSP70 inducibility is as relevant to life expectancy as telomeres. Therefore we are guaranteed at some stage in the near future to hear something about this genetic mechanism.

Regardless of what mechanism leads to the age-related decline in HSP70 inducibility, the most important consequence of this is the increase in activity of

JNK kinase, apoptosis and cell death (Volloch et al. 1998). Why did nature introduce this genetic mechanism, and why do cells always use this mechanism at the end of their life?

To better understand the suicidal behaviour of senescent cells consider yourself as such a cell. You had a rough life, you worked hard, and now you know that you have accumulated a lot of dangerous mutations and many damaged proteins. In principle you could still fight for your life: you still have your FORD machinery, battered, but still running, at your disposal; and you have stocks of ATP, albeit reduced, but still sufficient to ensure a minimum energy supply for the protein quality control system. Your co-chaperone partners and HDACs can still help you. However, you are a wise old cell; you certainly know that by prolonging your days, you will be passing dangerous pro-oncogenic mutations to your daughter cells, to your children, thereby increasing the likelihood of malignant tumour transformation of your "baby-cells" at each division event. So you take a courageous decision by pressing the "self-destruct button."

Thus, we can assume that a reduction of HSP70 inducibility in aging cells reflects the launch of a special genetic program (Fig. 7.3), aimed at activating JNK-dependent apoptosis and the destruction of old cells that have accumulated damaged proteins and dangerous mutations. Such a process(es) ensures the protection of the body.

References

Ambra R, Mocchegiani E, Giacconi R et al (2004) Characterization of the HSP70 response in lymphoblasts from aged and centenarian subjects and differential effects of in vitro zinc supplementation. Exp Gerontol 39(10):1475–1484

Ames BN, Shigenaga MK, Hagen TM (1993) Oxidants, antioxidants, and the degenerative diseases of aging. Proc Natl Acad Sci U S A 90:7915–7922

Batulan Z, Shinder GA, Minotti S et al (2003) High threshold for induction of the stress response in motor neurons is associated with failure to activate HSF1. J Neurosci 23(13):5789–5798

Beckman KB, Ames BN (1998) The free radical theory of aging matures. Physiol Rev 78:547–581

Bellmann K, Jäättelä M, Wissing D, Burkart V, Kolb H (1996) Heat shock protein HSP70 overexpression confers resistance against nitric oxide. FEBS Lett 391(1–2):185–188

Berlett BS, Stadtman ER (1997) Protein oxidation in aging, disease, and oxidative stress. J Biol Chem 272:20313–20316

Bhat RV, Budd Haeberlein SL, Avila J (2004) Glycogen synthase kinase 3: a drug target for CNS therapies. J Neurochem 89(6):1313–1317

Boyault C, Zhang Y, Fritah S et al (2007) HDAC6 controls major cell response pathways to cytotoxic accumulation of protein aggregates. Genes Dev 17:2172–2181

Corbi G, Conti V, Scapagnini G, Filippelli A, Ferrara N (2012) Role of sirtuins, calorie restriction and physical activity in aging. Front Biosci (Elite Ed) 4:768–778

Cortopassi GA, Wong A (1999) Mitochondria in organismal aging and degeneration. Biochim Biophys Acta 1410:183–193

Cosulich S, Clarke P (1996) Apoptosis: does stress kill? Curr Biol 6(12):1586–1588

Dall TM, Fulgoni VL 3rd, Zhang Y, Reimers KJ, Packard PT, Astwood JD (2009) Potential health benefits and medical cost savings from calorie, sodium, and saturated fat reductions in the American diet. Am J Health Promot 23(6):412–422

Dice JF (2007) Chaperone-mediated autophagy. Autophagy 3(4):295–299

Everitt AV, Le Couteur DG (2007) Life extension by calorie restriction in humans. Ann N Y Acad Sci 1114:428–433

Gabai VL, Kabakov AE (1993) Rise in heat-shock protein level confers tolerance to energy deprivation. FEBS Lett 327(3):247–250

Gabai VL, Meriin AB, Mosser DD et al (1997) HSP70 prevents activation of stress kinases. A novel pathway of cellular thermotolerance. J Biol Chem 272(29):18033–18037

Gabai VL, Meriin AB, Yaglom JA, Volloch VZ, Sherman MY (1998) Role of HSP70 in regulation of stress-kinase JNK: implications in apoptosis and aging. FEBS Lett 438(1–2):1–4

Gagliano N, Grizzi F, Annoni G (2007) Mechanisms of aging and liver functions. Dig Dis 25(2):118–123

Garrido C, Brunet M, Didelot C, Zermati Y, Schmitt E, Kroemer G (2006) Heat shock proteins 27 and 70: anti-apoptotic proteins with tumorigenic properties. Cell Cycle 5(22):2592–2601

Gutsmann-Conrad A, Heydari AR, You S, Richardson A (1998) The expression of heat shock protein 70 decreases with cellular senescence in vitro and in cells derived from young and old human subjects. Exp Cell Res 241(2):404–413

Hall DM, Oberley TD, Moseley PM et al (2000) Caloric restriction improves thermotolerance and reduces hyperthermia-induced cellular damage in old rats. FASEB J 14:78–86

Hands S, Sinadinos C, Wyttenbach A (2008) Polyglutamine gene function and dysfunction in the ageing brain. Biochim Biophys Acta 1779(8):507–521

Heydari AR, Takahashi R, Gutsmann A, You S (1994) HSP70 and aging. Experientia 50(11–12):1092–1098

Hsu AL, Murphy CT, Kenyon C (2003) Regulation of aging and age-related disease by DAF-16 and heat-shock factor. Science 300(5622):1142–1145

Hut HM, Kampinga HH, Sibon OC (2005) HSP70 protects mitotic cells against heat-induced centrosome damage and division abnormalities. Mol Biol Cell 16(8):3776–3785

Jäättelä M (1993) Overexpression of major heat shock protein HSP70 inhibits tumor necrosis factor-induced activation of phospholipase A2. J Immunol 151(8):4286–4294

Jin X, Wang R, Xiao C et al (2004) Serum and lymphocyte levels of heat shock protein 70 in aging: a study in the normal Chinese population. Cell Stress Chaperones 9(1):69–75

Johnson FB, Sinclair DA, Guarente L (1999) Molecular biology of aging. Cell 2:291–302

Kayani AC, Morton JP, McArdle A (2008) The exercise-induced stress response in skeletal muscle: failure during aging. Appl Physiol Nutr Metab 33(5):1033–1041

Kovacs JJ, Murphy PJ, Gaillard S et al (2005) HDAC6 regulates HSP90 acetylation and chaperone-dependent activation of glucocorticoid receptor. Mol Cell 18(5):601–607

Kregel KC (2002) Heat shock proteins: modifying factors in physiological stress responses and acquired thermotolerance. J Appl Physiol 92:2177–2186

Kregel KC, Moseley PL, Skidmore R, Gutierrez JA, Guerriero V (1995) HSP70 accumulation in tissues of heat-stressed rats is blunted with advancing age. J Appl Physiol 79:1673–1678

Li GC, Yang SH, Kim D et al (1995) Suppression of heat-induced HSP70 expression by the 70-kDa subunit of the human Ku autoantigen. Proc Natl Acad Sci U S A 92(10):4512–4516

Lindquist S, Craig EA (1988) The heat-shock proteins. Annu Rev Genet 22:631–677

Liu AY, Lee YK, Manalo D, Huang LE (1996) Attenuated heat shock transcriptional response in aging: molecular mechanism and implication in the biology of aging. EXS 77:393–408

Marques C, Guo W, Pereira P, Taylor A, Patterson C, Evans PC, Shang F (2006) The triage of damaged proteins: degradation by the ubiquitin-proteasome pathway or repair by molecular chaperones. FASEB J 20(6):741–743

Mayer MP, Bukau B (2005) HSP70 chaperones: cellular functions and molecular mechanism. Cell Mol Life Sci 62(6):670–684

McArdle A, Dillmann WH, Mestril R, Faulkner JA, Jackson MJ (2004) Overexpression of HSP70 in mouse skeletal muscle protects against muscle damage and age-related muscle dysfunction. FASEB J 18:355–357

Mignotte B, Vayssiere JL (1998) Mitochondria and apoptosis. Eur J Biochem 252(1):1–15

Min JN, Whaley RA, Sharpless NE, Lockyer P, Portbury AL, Patterson C (2008) CHIP deficiency decreases longevity, with accelerated aging phenotypes accompanied by altered protein quality control. Mol Cell Biol 28(12):4018–4025

Minois N, Khazaeli AA, Curtsinger JW (2001) Locomotor activity as a function of age and life span in *Drosophila melanogaster* overexpressing HSP70. Exp Gerontol 36(7):1137–1153

Mosser DD, Martin LH (1992) Induced thermotolerance to apoptosis in a human T lymphocyte cell line. J Cell Physiol 151(3):561–570

Mosser DD, Caron AW, Bourget L, Denis-Larose C, Massie B (1997) Role of the human heat shock protein HSP70 in protection against stress-induced apoptosis. Mol Cell Biol 17(9):5317–5327

Njemini R, Abeele MV, Demanet C, Lambert M, Vandebosch S, Mets T (2002) Age-related decrease in the inducibility of heat-shock protein 70 in human peripheral blood mononuclear cells. J Clin Immunol 22(4):195–205

Papaconstantinou J (1994) Unifying model of the programmed (intrinsic) and stochastic (extrinsic) theories of aging. Ann NY Acad Sci 719:195–211

Powis G, Briehl M, Oblong J (1995) Redox signalling and the control of cell growth and death. Pharmacol Ther 68:149–173

Raffray M, Cohen GM (1997) Apoptosis and necrosis in toxicology: a continuum or distinct modes of cell death? Pharmacol Ther 75(3):153–177

Rujano MA, Bosveld F, Salomons FA et al. (2006) Polarised asymmetric inheritance of accumulated protein damage in higher eukaryotes. PLoS Biol 4(12):e417

Salvesen GS, Dixit VM (1997) Caspases: intracellular signaling by proteolysis. Cell 91(4):443–446

Seimiya H, Mashima T, Toho M, Tsuruo T (1997) c-Jun NH2-terminal kinase-mediated activation of interleukin-1beta converting enzyme/CED-3-like protease during anticancer drug-induced apoptosis. J Biol Chem 272(7):4631–4636

Sherman MY, Goldberg AL (2001) Cellular defenses against unfolded proteins: a cell biologist thinks about neurodegenerative diseases. Neuron 29(1):15–32

Sherman H, Frumin I, Gutman R, Chapnik N, Lorentz A, Meylan J, le Coutre J, Froy O (2011) Long-term restricted feeding alters circadian expression and reduces the level of inflammatory and disease markers. J Cell Mol Med 15(12):2745–2759

Simon MM, Reikerstorfer A, Schwarz A et al (1995) Heat shock protein 70 overexpression affects the response to ultraviolet light in murine fibroblasts. Evidence for increased cell viability and suppression of cytokine release. J Clin Invest 95(3):926–933

Singh R, Kølvraa S, Bross P et al (2006) Reduced heat shock response in human mononuclear cells during aging and its association with polymorphisms in HSP70 genes. Cell Stress Chaperones 11(3):208–215

Singh R, Kolvraa S, Rattan SI (2007) Genetics of human longevity with emphasis on the relevance of HSP70 as candidate genes. Front Biosci 12:4504–4513

Steinkraus KA, Smith ED, Davis C et al (2008) Dietary restriction suppresses proteotoxicity and enhances longevity by an hsf-1-dependent mechanism in *Caenorhabditis elegans*. Aging Cell 7(3):394–404

Verheij M, Bose R, Lin XH, Yao B et al (1996) Requirement for ceramide-initiated SAPK/JNK signalling in stress-induced apoptosis. Nature 380(6569):75–79

Visala Rao D, Boyle GM, Parsons PG, Watson K, Jones GL (2003) Influence of ageing, heat shock treatment and in vivo total antioxidant status on gene-expression profile and protein synthesis in human peripheral lymphocytes. Mech Ageing Dev 124(1):55–69

Volloch V, Mosser DD, Massie B, Sherman MY (1998) Reduced thermotolerance in aged cells results from a loss of an HSP72-mediated control of JNK signaling pathway. Cell Stress Chaperones 3(4):265–271

Webb SJ, Harrison DJ, Wyllie AH (1997) Apoptosis: an overview of the process and its relevance in disease. Adv Pharmacol 41:1–34

Westerheide SD, Anckar J, Stevens SM Jr, Sistonen L, Morimoto RI (2009) Stress-inducible regulation of heat shock factor 1 by the deacetylase SIRT1. Science 323(5917):1063–1066

Westphal CH, Dipp MA, Guarente L (2007) A therapeutic role for sirtuins in diseases of aging? Trends Biochem Sci 32(12):555–560

Winklhofer KF, Tatzelt J, Haass C (2008) The two faces of protein misfolding: gain- and loss-of-function in neurodegenerative diseases. EMBO J 27(2):336–349

Xia Z, Dickens M, Raingeaud J, Davis RJ, Greenberg ME (1995) Opposing effects of ERK and JNK-p38 MAP kinases on apoptosis. Science 270(5240):1326–1331

Zanke BW, Boudreau K, Rubie E et al (1996) The stress-activated protein kinase pathway mediates cell death following injury induced by cis-platinum, UV irradiation or heat. Curr Biol 6(5):606–613

Chapter 8
The Role of HSP70 in the Protection of: (A) The Brain in Alzheimer's Disease and (B) The Heart in Cardiac Surgery

Abstract The accumulation of aggregated, misfolded proteins and the appearance of neurotoxic aggregates of Aβ and tau proteins play a key role in the development of Alzheimer's disease. HSP70 can inhibit neurodegeneration associated with Alzheimer's disease because this protein can: (i) aid in the degradation of intracellular and extracellular Aβ aggregates; (ii) restrict tau protein hyperphosphorylation and facilitate the degradation of dysfunctional tau proteins; (iii) limit NO overproduction; and (iv) regulate apoptosis. It is also likely that HSP70 may delay the development of Alzheimer's disease by limiting insulin receptor desensitization. HSP70 can limit ischemia myocardial injury by: (i) maintaining protein homeostasis in cells; (ii) stabilizing lysosomal membranes; (iii) inhibiting the excessive activation of ADP-ribose polymerase; and (iv) blocking ischemia-induced apoptosis. During the excessive systemic inflammatory response syndrome (SIRS) that occurs in heart surgery, extracellular HSP70 initiates inflammatory effects through the stimulation of immune cell receptors. In contrast, intracellular HSP70, exerts anti-inflammatory effects on the inflammatory balance of SIRS by inhibiting proinflammatory signaling in immune cells.

Keywords HSP70 • Alzheimer disease • Insulin • Ischemia • Systemic inflammatory response syndrome

8.1 The Role of HSP70 in the Protection of the Brain in Alzheimer's Disease

In the previous chapter, I discussed what happens to HSPs, and more generally with the proteins quality control system (FORD machinery) in a senescent cell and in an aging body. Briefly, with age there is a decrease in the inducibility and effectiveness of the cell chaperone system. Globally, this leads to two consequences. First, the damaged, denatured proteins and toxic protein aggregates increase in number, and, second, HSPs stop monitoring two genetic programs: the program of cell replicative aging and the program of apoptosis. As a result, the cell stops dividing and dies. Physiological cessation of division and death of senescent cells are fully justified. It protects the body from mitotic proliferation of potentially dangerous prooncogenic and other mutations acquired during the life of a cell.

I. Malyshev, *Immunity, Tumors and Aging: The Role of HSP70*,
SpringerBriefs in Biochemistry and Molecular Biology,
DOI: 10.1007/978-94-007-5943-5_8, © The Author(s) 2013

The disruption of normal physiological aging of cells, which may be caused by various genetic factors and adverse environmental stresses, can lead to the development of age-related diseases. In this chapter I will talk about one of the most widely spread diseases in the world, Alzheimer's disease, and the role heat shock proteins, especially HSP70, play in this disease. However, let me introduce Alzheimer's disease and the issues surrounding the link between HSPs and this disease.

8.1.1 Pathogenesis of Alzheimer's Disease: The Disease is Still Incurable, so There Must be Something Really Important that Remains Unknown

A whole army of scientists and physicians have been studying Alzheimer's disease for over a hundred years in an effort to find ways to treat and prevent this disease. Every time we open a new issue of a scientific journal, we often discover something new, interesting and important about the molecular mechanisms of neurodegeneration and dementia.

Alzheimer's is a neurodegenerative disease of the central nervous system. One in three people (Sadik and Wilcock 2003) who live to an old age are likely to be affected. Starting subtly and stealthily, the gradual progression of Alzheimer's disease leaves no chance for recovery. Alzheimer's disease, long before the biological death of the body, gradually robs a person of the most important thing that makes us *Homo sapiens*, our memory. As a result of this disease, we lose the ability to establish causal relationships, to perceive and analyze new information, and to recognize friends and relatives.

What phenotypic features appear to the brains of patients with Alzheimer's? Genetic analysis of families presenting Alzheimer's dementia were found to have mutations in the amyloid precursor protein (APP) and presenilin 1 (PS1) (Mullan et al. 1992; Van Broeckhoven et al. 1990; Goate et al. 1991; Chartier-Harlin et al. 1991; Campion et al. 1995; Perez-Tur et al. 1996; Sherrington et al. 1995) (Fig. 8.1).

In healthy humans, APP is cleaved by the gamma-secretase complex, resulting in the formation of an amyloid peptide in the alpha form (Aα). The central component of the gamma-secretase complex is PS1. Aα incorporates into the membrane and performs important physiological functions. As we age the gamma secretase complex begins making mistakes because of mutations to the APP and/or PS1 genes. The secretase cleaves the APP at a slightly different position. The error is only a few amino-acid residues, but the consequences are dramatic. As a result, the neurotoxic, beta form of the amyloid protein (Aβ) appears in hippocampal neurons and the cerebral cortex.

Aβ may remain in the nerve cell or enter into the extracellular space (Hsiao et al. 1996; Duff et al. 1996; Holcomb et al. 1998, 1999; Takeuchi et al. 2000; Borchelt et al. 1997). Aβ has a high tendency to form neurotoxic oligomers and multimeric aggregates (Golde et al. 1992; Estus et al. 1992; Shoji et al. 1992; Halverson et al. 1990; Meyer-Luehmann et al. 2008; Spires-Jones et al. 2008; Shankar et al. 2007, 2008; Walsh et al. 2002; Cleary et al. 2005; Gong et al. 2003; Caughey and Lansbury 2003).

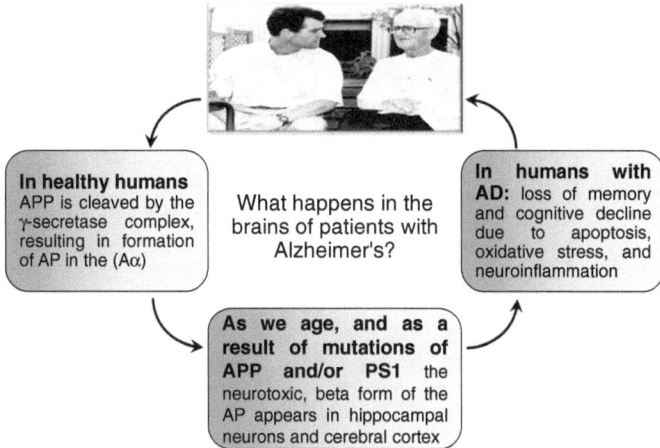

Fig. 8.1   What happens in the brains of patients with Alzheimer's

There is ample evidence that Aβ plays an important role in the pathogenesis of Alzheimer's disease. Neurotoxicity of Aβ is due to its ability to induce apoptosis, oxidative stress, NO overproduction (Christen 2000; Clippingdale et al. 2001) and neuro-inflammation (Sasaki et al. 1997; Hu et al. 1998; Stalder et al. 2005; Simard et al. 2006). Together, these processes lead to the death of neurons, atrophy of some brain regions with cognitive decline evident as loss of memory and the inability to recognize close friends and family members.

Despite the well-proven role of Aβ in the pathogenesis of Alzheimer's disease, the use of therapies aimed at reducing Aβ accumulation or production of this peptide has had limited clinical success (Schenk et al. 2005; Hock et al. 2003; Nitsch et al. 2008; Wilcock et al. 2004, 2006; Osborne 2008). It is therefore evident that apart from Aβ, there are other factors that play an important role in the pathogenesis of this disease.

One of these factors proved to be a mutation in the tau protein gene (Hutton et al. 1998; Spillantini et al. 1998; Rizzini et al. 2000) (Fig. 8.2). The normal tau protein binds to microtubules and regulates their polymerization and stability. Mutant forms of the tau protein, as opposed to the normal ones, are phosphorylated by GSK3β and Cdk5 kinases (Noble et al. 2005; Kosik et al. 2002). Tau hyperphosphorylation triggers aggregation of this protein and formation of intracellular neurotoxic fibrillar tau tangles. It was shown that this process is strongly correlated with the death of neurons and cognitive impairment (Santacruz et al. 2005; Ramsden et al. 2005). The ideas about the pathogenic role of abnormal tau protein were immediately supported by initial clinical research, which found that an inhibitor of tau protein aggregation significantly improved the cognitive function in Alzheimer's patients (Opar 2008). Thus it is currently known that the accumulation of aggregated, misfolded proteins and the appearance of abnormal oligomeric forms and neurotoxic aggregates of Aβ and tau protein, as well as mutant PS1 proteins play a key role in the development of Alzheimer's disease.

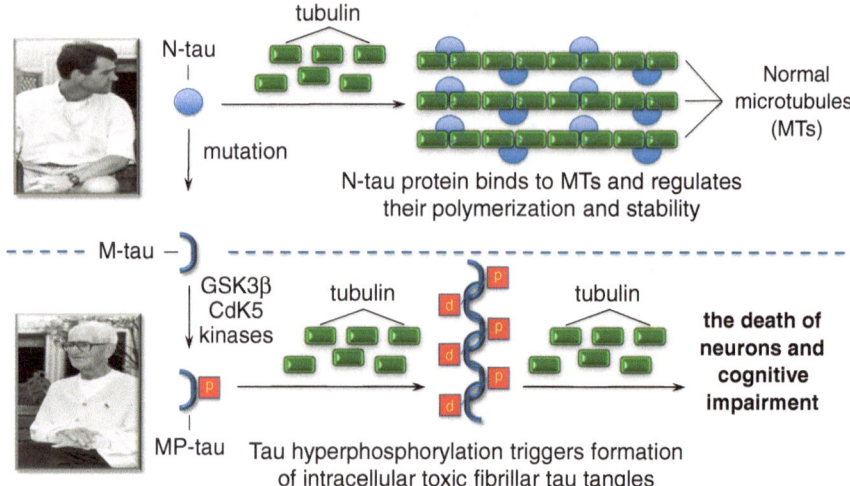

Fig. 8.2 A mutation in the tau protein gene plays an important role in the pathogenesis of Alzheimer's disease

8.1.2 The Role of HSP70 in Alzheimer's Disease. Sometimes You Feel Like You are Balancing on a Sword Blade

The realization that protein homeostasis disorder and the appearance of mutant proteins and damaged forms of certain proteins played key roles in neurodegeneration immediately gave rise to two hypotheses.

The first hypothesis proposed that a disturbance of protein metabolism could be associated with the age-related weakening of the chaperone protein quality control system. This hypothesis was supported in studies that showed that the content of HSP70 mRNA in patients with Alzheimer's disease was significantly lower than that in old healthy people of the same age (Wakutani et al. 1995; Getchell et al. 1996). In addition, it was found that abnormality in hippocampal neurons led to cognitive decline and Alzheimer's disease. Moreover, such neurons also had low basal levels of HSP70 when compared with other brain regions (Chen and Brown 2007). The hippocampus is responsible for cognitive functions of the brain and memory. The question arises then: why did nature not arrange for the protection of this region of the brain? A higher basal level of HSP70 could provide "pre-protection" of these neurons from stress-induced disturbances in protein homeostasis and ensure protection of the brain cognitive functions from detrimental effects from everyday stresses. This question, like many others related to the issue of neurodegeneration, remains unanswered. One hypothetical explanation put forward is that high basal levels of HSP70 somehow interfere with cognitive functions.

The second hypothesis was that HSPs can inhibit neurodegeneration in Alzheimer's disease (Renkawek et al. 1993; Pappolla et al. 1996; Brown and

Gozes 1998; Ohtsuka and Suzuki 2000). Indeed, it was found that, despite a general age-related decline of the chaperone system inducibility in the brain of Alzheimer's patients, the levels of heat shock proteins HSP27, HSP70 and CHIP in the regions affected by the pathological process were increased (Yoo et al. 1999; Renkawek et al. 1994; Petrucelli et al. 2004). Experiments with increased levels of expression of HSP70 showed that HSP70 protects neurons from neuro-degenerative processes (Magrane et al. 2004; Smith et al. 2005). We can list at least six mechanisms of the neuroprotective effect of HSP70: (1) disaggregation and degradation of intracellular Aβ aggregates; (2) disaggregation of extracellular Aβ aggregates; (3) an increase in the elimination of Aβ from intercellular spaces; (4) restriction of tau protein hyperphosphorylation, disaggregation and degradation of abnormal tau proteins; (5) limiting NO overproduction; and (6) restriction of apoptosis.

Mechanism 1. Studies using cell cultures and the human brain have shown that Aβ can accumulate inside nerve cells, either by intracellular production, or by the uptake of extracellular Aβ (Yang et al. 1998; Gouras et al. 2000; Sun et al. 2002). Intracellular Aβ neurotoxicity may manifest itself long before the extracellular accumulation of Aβ (Kumar-Singh et al. 2000; Wirths et al. 2001). Moreover, after lysis of nerve cells, intracellular Aβ contributes significantly to the formation of senile plaques (D'Andrea et al. 2001). Different research has shown that intracellular Aβ forms complexes with several proteins, many of which have chaperone activity (Yan et al. 1997; Fonte et al. 2002; Cottrell et al. 2005). Intracellular Aβ was found to interact with HSP70 in a specific manner (Fonte et al. 2002). (Fonte et al. 2002; Evans et al. 2006) found that HSP70 can inhibit the earliest steps of aggregation of intracellular Aβ, thereby completely suppressing its neurotoxicity (Magrané et al. 2004; Ansar et al. 2007) and the formation of the degenerative neuron phenotype. These data suggest that intracellular Aβ is recognized by the chaperone system of the cell, including HSP70, as an abnormal protein. As a result, HSP70, together with other chaperone proteins, rearranges Aβ metabolism from the secretory pathway and the formation of extracellular plaques to an alternative pathway of Aβ disaggregation and degradation (Fig. 8.3).

Mechanism 2. The extracellular accumulation of beta-amyloid plaques is believed to be one of the most important triggers of neurodegenerative processes. It was shown that low molecular weight stress proteins may prevent the formation of Aβ aggregates in vitro, and that extracellular HSP70 and HSP90 can disaggregate toxic amyloid aggregates (Koren et al. 2009) (Fig. 8.4).

Mechanism 3. The content of Aβ in the extracellular space is determined by the ratio between the rates of its production and elimination. Kakimura et al. (2002) showed that extracellular HSP90, HSP70 and HSP32 may actively influence this ratio through activation of microglia and the induction of IL-6 and TNF-α production. Microglial cytokines enhance phagocytosis, thus contributing to the removal of neurotoxic Aβ (Fig. 8.5). However, an excessive increase in the production of proinflammatory cytokines may enhance neuro-inflammation. Therefore, once again, a particular balance of the effects of cytokines on HSPs should be observed.

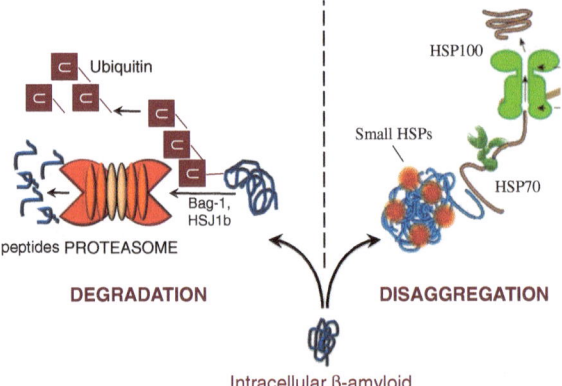

Fig. 8.3 HSP70 rearranges Aβ metabolism from the secretory pathway and the formation of extracellular plaques to an alternative pathway of Aβ disaggregation and degradation

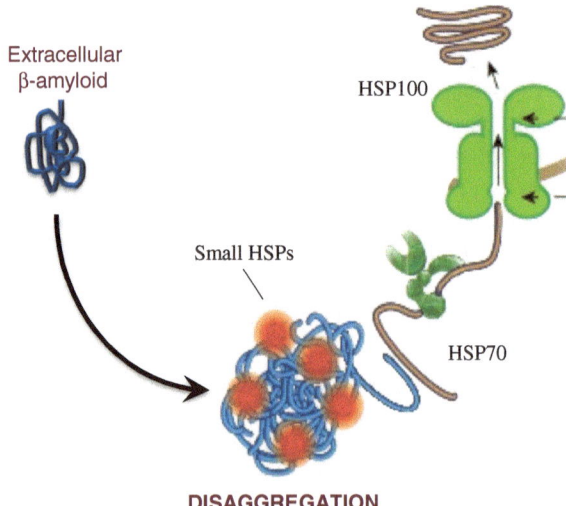

Fig. 8.4 Small stress proteins may prevent the formation of Aβ aggregates, and extracellular HSP70 and HSP90 can disaggregate toxic amyloid aggregates

Mechanism 4. Normal Tau is a protein that promotes polymerization and stabilization of microtubules. However, hyperphosphorylation of the tau protein is the major trigger for the formation of fibrillar neuronal tangles in neurons affected by Alzheimer's disease. Thus, the interaction of tau with microtubules is disrupted, thereby disturbing polymerisation and the stability of microtubules (Weaver et al. 2000; Guillozet-Bongaarts et al. 2005; Pickering-Brown et al. 2000). Johnson et al. (1993) discovered that HSP70 forms a stable complex with the tau protein. Subsequently, Kirby et al. (1994) proved that HSP70 binding to tau protects tau from hyperphosphorylation and, therefore, limits the formation of neurofibrillary tangles (Fig. 8.6). In contrast, it was shown that reducing HSP70 levels leads to the formation of neurofibrillar tangles (Dou et al. 2003).

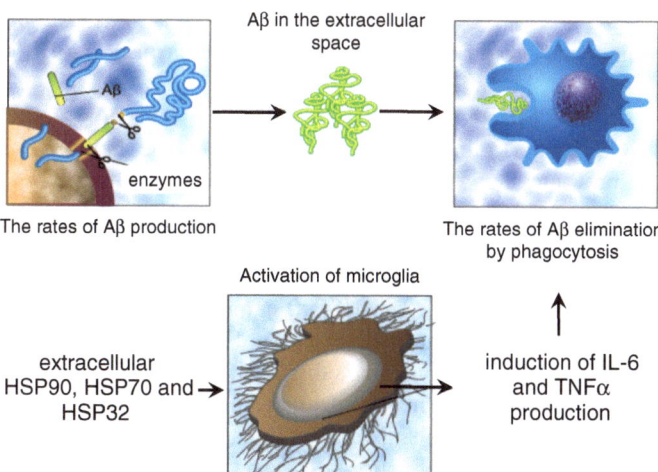

Fig. 8.5 HSPs induced microglial cytokines enhance phagocytosis, thus contributing to the removal of neurotoxic Aβ

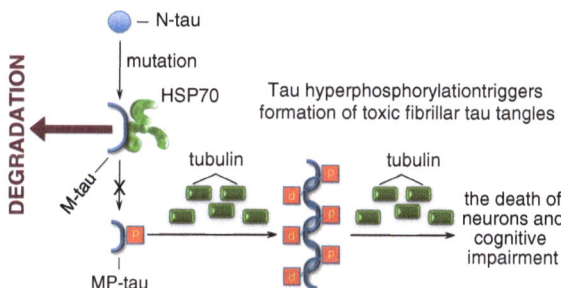

Fig. 8.6 HSP70 binding to tau protects tau from hyperphosphorylation and, therefore, limits the formation of neurofibrillary tangles, as well as reduce mutant tau intracellular concentration by targeting these proteins for degradation

In addition, the chaperone proteins such as HSP27, HSP70 and CHIP can detect abnormal tau and reduce its intracellular concentration by targeting this protein for degradation in proteasomes or lysosomes (Petrucelli et al. 2004; Dou et al. 2003; Shimura et al. 2004; Sarkar et al. 2008; Elliott et al. 2007). Taken together, this data suggests that chaperones are required to maintain tau in the non-aggregated, normal state.

Other studies have shed light on what mechanism the chaperone system uses to control abnormally accumulated tau. Protein kinase Akt, which has elevated levels in Alzheimer's patients, plays a very important role in this mechanism (Pei et al. 2003). Akt is a kinase that, on the one hand, can phosphorylate the tau protein (Pei et al. 2003) thus stimulating its aggregation, and, on the other hand, prevent tau ubiquitination and the subsequent degradation of abnormal tau (Dickey et al. 2006, 2008). A tentative model that illustrates our current understanding of tau protein metabolism under the control of the HSP70/HSP90 chaperone system is shown in Fig. 8.7.

Kinases, such as MARK2 and Akt phosphorylate a specific area of the tau protein (KXGS motifs); this phosphorylation, with the help of the chaperone system, prevents the degradation of abnormal tau. Another abnormal or misfolded tau, non-phosphorylated at the KXGS motif, is recognized by the HSP40/HSP70 complex and is sent directly to the proteasome for degradation, or through the formation of a complex with HSP90. Akt can recognize and bind the HSP40/HSP70/HSP90 complex. The HSP90 complex can provide refolding and recovery of tau, or it can send tau to degradation. However, in the presence of Akt, the degradation of tau by this complex is disrupted and tau accumulates. Note that Akt is also an HSP90 client and may be degraded along with the tau protein.

Mechanism 5. The next mechanism of neuronal death in Alzheimer's disease is associated with NO overproduction. NO overproduction plays an important pathogenic role in inflammatory reactions during Alzheimer's disease. The brain contains all three NOS isoforms. Neuronal NOS is expressed throughout the brain, including cerebellum, cortex, hippocampus, amygdala and substantia nigra (Vincent and Kimura 1992). Endothelial NOS is localized mainly in endothelial cells. Inducible NOS is absent in healthy brain tissue, but can be expressed in cells following brain injury. NO overproduction can be associated with excessive activation of iNOS by cytokines in microglial cells and astrocytes. NO causes

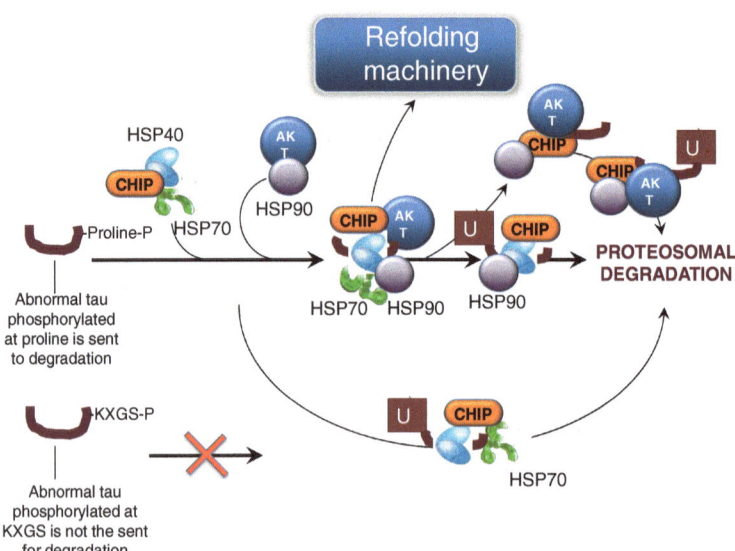

Fig. 8.7 A tentative model that illustrates tau protein metabolism under the control of the HSP70/HSP90 chaperone system and protein kinase Akt

neuronal death mainly by damaging mitochondria and triggering apoptosis (Charbrier et al. 1999).

It was shown that HSP70 may limit NO production and thus reduce neuro-inflammation (Calabrese et al. 2000) by inhibiting the expression of cytokine and iNOS genes (Yoo et al. 2000a, b). This effect is due to the ability of intracellular HSP70 to block activation of NFkB, the main pro-inflammatory transcription factor (Fig. 8.8).

Mechanism 6. Neuronal death in Alzheimer's disease occurs mainly through the apoptotic pathway. Neuronal apoptosis can be triggered by neurotoxic forms of Aβ, tau-protein, NO overproduction and a variety of stressors, the sensitivity to which is significantly increased in aging neurons. In the previous chapters, I have already examined how HSP70 may limit apoptosis in cells. It is possible that HSP70 may limit apoptosis in neurons using the same mechanisms (Fig. 3.4). Briefly, HSP70 may block apoptosis by inhibiting mitochondrial release of the proapototic factors, cytochrome c and Araf-1, inhibit the activity of AIF, caspase-9 and JNK, as well as by increasing the level of Bcl-2 and reducing the level of Bax. In general, we can conclude that HSP70 is an important endogenous system for protection of hippocampal neurons against Alzheimer's disease.

There is an interesting and important point to the story describing the intracellular relationship between HSPs and Aβ (Fig. 8.9). HSE, the site that can bind HSF1, was discovered in the promoter of the *app* gene (Salbaum et al. 1998). That means that under stress, when HSF1 is activated, this transcription factor can increase the expression of both *HSP* genes, such as *HSP27, HSP40, HSP70* and *HSP90*, and the *app* gene (Dewji 2005, 2006; Dewji et al. 1995). These data,

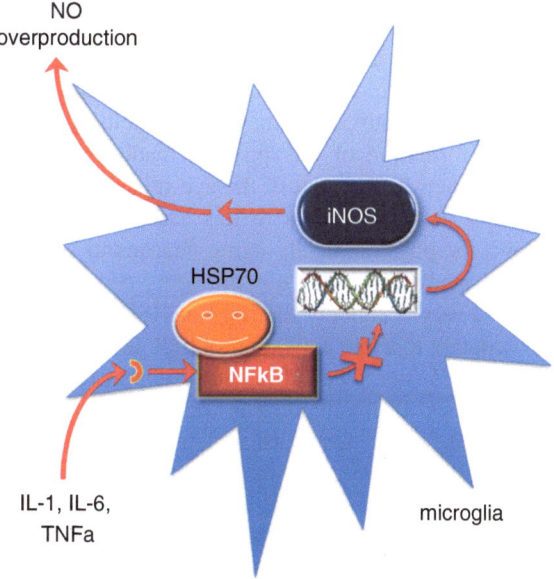

Fig. 8.8 HSP70 may limit NO production and thus reduce neuro-inflammation by inhibiting the expression of cytokine and iNOS genes

Fig. 8.9 The intracellular
relationship between HSF-1
and Aβ: a positive feedback
mechanism, which would
increase the production of
neurotoxic Aβ, may form in
cells of an aging brain

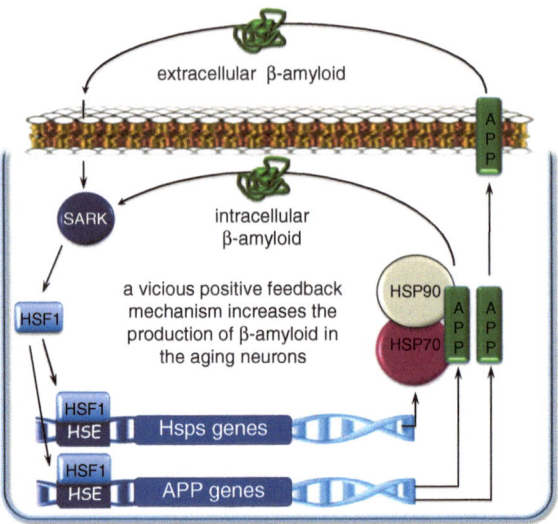

on the one hand, allow us to understand and explain the molecular mechanism by
which stress can contribute to the development of Alzheimer's disease, and, on
the other hand, offer a hypothesis about why the risk of early neurodegenerative
changes and Alzheimer's disease increases with age.

It could be that during aging, the HSF1 affinity to the promoter of the *HSP70*
gene reduces, while the affinity to the promoter of the *app* gene increases. The
increase in the synthesis of APP and the decrease in HSP70 inducibility would
then be associated with an aging brain. Consequently, such protein production regulation may, under certain stress conditions, contribute to the formation of neurotoxic Aβ within and outside the cell. Extracellular Aβ can activate stress-induced
kinases, which in turn can phosphorylate and thus activate HSF1 (Koren et al.
2009). Therefore, a positive feedback mechanism, which would increase the production of neurotoxic Aβ, may form in cells of an aging brain.

The formation of this positive feedback mechanism, which involves HSF1,
allows a new assessment of the age-related decrease in HSF1 activity in a healthy
brain. It is possible, that in this case, the age-related decline in HSF1 activity
allows the physiologically aging neurons to substantially slow down the production of neurotoxic Aβ. Thus the aging neuron, if it wants to stay healthy, has to
balance very accurately the level of HSF1 activity. Here, HSF1 activity should be
low enough to avoid activation of high levels of amyloid production, yet the activity of HSF1 should be sufficient to support HSP70 inducibility to maintain protein
homeostasis. If this hypothesis was confirmed, HSF1 would potentially become a
very good "target" for pharmacological treatment of Alzheimer's disease.

Once again, we can only regret that hippocampal neurons contain low basal
levels of protective HSP70, as compared to other regions of the brain and other
cells. I have already mentioned that this is probably due to the fact that high levels
of HSP70 in some way impede the execution of cognitive functions.

An amazing thing happens! When neurons are damaged, the surrounding microglial cells and astrocytes begin secreting high-levels of HSP70. These HSP70 molecules, besides participating in the disaggregation of extracellular Aβ aggregates, are also taken up by neurons and begin protecting neurons. Neurons, not having sufficient number of their own HSP70 use HSP70 released from neighboring glial cells and astrocytes (Tytell et al. 1986). A stunning example of mutual help among brain cells!

8.1.3 The Insulin Hypothesis of Alzheimer's Disease and a Possible Role of HSP70

In analyzing the role of HSPs in Alzheimer's disease, we were primarily focused on the prevailing hypothesis, formed over many years, about this disease involving the aggregation of proteins and impaired homeostasis of Aβ and the tau protein.

However, recently some truly revolutionary discoveries have been made, and they can radically alter our views on the etiology of Alzheimer's disease. If in 2009 you would have opened the BMB reports magazine, you would have found an interesting article there: "Insulin resistance and Alzheimer's disease". The author, Suzanne de la Monte (de la Monte 2009), argued that the earliest event that triggers neurodegeneration and development of Alzheimer's disease is a reduction in insulin levels and insulin receptor desensitization in the central nervous system. Both factors were found in the brains of patients with Alzheimer's disease, and both of them, as experimentally shown, are likely to disrupt insulin-dependent signalling pathways and inhibit insulin-dependent genes in brain cells. According to Suzanne de la Monte that is precisely what leads to the development of the whole spectrum of Alzheimer's disease symptoms, namely, accumulation of Aβ oligomers and aggregates, neurofibrillar tangles of hyperphosphorylated tau protein, oxidative stress, activation of apoptosis and neuronal death.

The most amazing thing is that insulin resistance in the brain can develop independently, whereas the insulin sensitivity of peripheral cells of other organs is not disturbed, that is, in the absence of type-2 diabetes. This observation gave grounds to Suzanne de La Monte to propose the existence of a specific cerebral form of "type-3 diabetes mellitus", which is the cause of Alzheimer's disease (Lester-Coll et al. 2006). This hypothesis has been supported by studies where intranasal administration of insulin significantly improved the cognitive function without changing peripheral glucose metabolism (Reger et al. 2008).

The insulin hypothesis, however, does not negate a number of cases when an increase in the level of HSP70 or one of its co-chaperones helped to overcome neurodegeneration symptoms (Bonini 2002; Klucken et al. 2004). However, if the insulin hypothesis of Alzheimer's disease is confirmed, we should be prepared to revise the role of HSPs as well. There is a lot of data, obtained on different types of cells, which suggest that HSP70 may play a specific role in the insulin-dependent signaling and possibly in the development of Type-3 diabetes and Alzheimer's disease. All these data, obtained mainly on hepatic cells, suggests that HSP70 may limit insulin resistance in cells as well as disorders of insulin-dependent signaling (McCarty 2005; Zachayus et al. 1996; Morino et al. 2008; Marucci et al. 2009).

However, once again, these data were obtained mainly on hepatic cells; thus we should cautiously extrapolate the data to what happens in the brain. One of the first studies on the relationship between HSP70 and insulin-dependent mechanisms in brain neurons was done by the Piccoletti group in Italy (Piccoletti et al. 2001). This study showed that the activation of the HSF1 transcription factor and the accumulation of HSP70 in neurons are accompanied by activation of insulin receptors. Thus, it can be suggested with certain caution that HSP70 may delay the development of neurodegeneration in Alzheimer's disease by limiting insulin receptor desensetization. Once again; however, this hypotesis requires serious experimental verification.

8.2 The Role of HSP70 in Protecting the Heart in Heart Surgery

We discussed how HSPs protect our brain, our cognitive functions and our consciousness. Now I turn to the role these proteins play in protecting the heart, which is, as previously believed, a "container for our soul".

When we hear, "he has a bad heart," our own heart shrinks with pain, especially when it is about someone close to us. Since the heart is not a paired organ, it cannot be removed like a dysfunctional kidney, and thus, heart-related diseases are often serious illnesses. Unlike the stomach and even the brain, removal of a piece of a damaged heart is not a viable option. The heart is the only organ that operates on the "all or nothing" principle, and thus a sick heart represents a serious health risk! Therefore, heart surgery is one of the best equipped and high-tech areas in medicine.

Sixty years ago, in April 1951, Dr. Dennis and his team from the University of Minnesota hospital made a huge technological breakthrough in heart surgery. They were the first to carry out an operation on an open heart by stopping the heart using the "iron heart" machine. Since then, many heart diseases have been treated and are not considered a sure death sentence. These days this device is called the artificial circulation machine, or "heart–lung machine", and the operations are known as cardio-pulmonary bypasses. Without this device it would not be possible to carry out operations that are often the last hope for survival of cardiological patients. These operations are coronary artery bypass surgery, cardiac valve repair and/or replacement (aortic valve, mitral valve, tricuspid valve, pulmonic valve), repair of large septal defects (atrial septal defect, ventricular septal defect, atrioventricular septal defect), repair and/or palliation of congenital heart defects (Tetralogy of Fallot, transposition of the great vessels), transplantation (heart transplantation, lung transplantation, heart–lung transplantation), repair of some large aneurysms (aortic aneurysms, cerebral aneurysms), pulmonary thromboendarterectomy and pulmonary thrombectomy.

The above list is rather extensive. A lot of terminally ill patients have been saved over the last half-a-century by using the artificial circulation system; however, some serious complications were also discovered. The most serious ones are ischemia and reperfusion injury, and the excessive systemic inflammatory response syndrome (SIRS). These two complications often lead to irreversible cardiac dysfunction, multiple organ failure and death.

In studying these complications, HSP70 proteins have attracted particular attention. However, before answering the question why HSP70 attracted the attention of scientists and cardiologists, it is important to say a few words about what these complications are and why they occur.

8.2.1 Ischemic Complications Associated with Open Heart Surgery, or How We are Hoisted by Our Own Petard

We must say at once that in open heart surgery, many of which are carried out to eliminate myocardial ischemia, it is not possible to avoid further ischemic damage to cardiomyocytes. Clamping the aorta, connecting the extracorporal circulation apparatus, and stopping the heart, lead directly to myocardial ischemia during surgery because the heart has to stop functioning for a period of time. In addition, restoration of blood supply to the heart after the surgery and reoxygenation of the ischemic tissue inevitably leads to oxidative stress and further reperfusion injury. During ischemia and subsequent reperfusion, the myocardium experiences denaturation of proteins and the development of all three classic types of cellular damage. I discussed these points in Chap. 3, namely, hypoxic damage, free radical damage and calcium overload. These injuries seriously contribute to heart rhythm disturbances and the development of postoperative atrial fibrillation (Archbold and Curzen 2003; St Rammos et al. 2002).

Reoxygenation of ischemic tissue also triggers local inflammation and the release of proinflammatory cytokines into the blood. These cytokines act on endothelial cells of coronary vessels and increase the production of adhesion molecules by these cells. The appearance of adhesion molecules on the surface of the endothelium contributes to the attachment of activated neutrophils and monocytes to the endothelium, and subsequent migration of these cells into the interstitial space of reperfused myocardium. There, the activated neutrophils release aggressive reactive oxygen species (ROS). Unfortunately the neutrophils do not care which cells are attacked by ROS; pathogens that must be destroyed or cardiomyocites that had just survived ischemia. The activated neutrophils attack all living cells in their path, provoking their death by necrotic or apoptotic processes.

8.2.2 The Role of HSP70 in Protecting the Heart From Ischemia/Reperfusion Injury During Heart Surgery

The development of ischemic heart damage is associated with hypoxia, activation of free radical processes, calcium overload, and protein denaturation. All these factors have been shown to be involved in the activation of HSP70 synthesis during ischemia. The value of ischemia-induced HSP70 synthesis has become clear in the study of the ischemic preconditioning phenomenon. The essence of this phenomenon lies in the fact that brief ischemia, which induces a small damage and accumulation of HSP70 in the heart, results in an increase of myocardial resistance to subsequent severe, prolonged ischemia and subsequent reperfusion (Taggart et al. 1997; Demidov et al. 1999; Giannessi et al. 2003). The conclusion from these studies was that the activation of HSP70 by damaging factors forms a basis for auto-adaptation of the myocardium to injury.

The protective role of HSP70 in the heart has been repeatedly confirmed in various experiments (Jayakumar et al. 2000, 2001; Okubo et al. 2001; Currie et al. 1998). It was shown, in particular, that after surgery the heart contractile function in patients, who had had higher myocardial levels of HSP70 before surgery, recovered faster than patients with lower levels of this protein (Demidov et al. 1999; Giannessi et al. 2003).

Naturally, once the high effectiveness of HSP70 in cardioprotection was shown there was significant interest in defining what and how HSP70 functions as a cardioprotectant. As a result, it was discovered that HSP70 can limit ischemia/reperfusion myocardial injury using at least four mechanisms: (1) by maintaining the protein homeostasis in cells; (2) by stabilizing lysosomal membranes; (3) by inhibiting the excessive activation of ADP-ribose polymerase; and (4) by blocking ischemia-induced apoptosis.

The first defensive mechanism is connected with HSP70 ability to maintain protein homeostasis in damaged ischemic and reperfused cells (Gabai and Kabakov 1993; Kabakov and Gabai 1995). I have presented this mechanism in detail in the third chapter. The main points of the mechanism are:

1. To maintain protein homeostasis in a damaged cell, the HSP70-dependent protein quality control system, FORD, prevents protein aggregation, disaggregates formed aggregates and degrades "incorrigible" proteins.
2. The restoration of protein homeostasis forms a specific cell defense from hypoxic injury due to HIF-1 activation; the defense from radical injury—by increasing the activity of antioxidants, and from calcium overload—by reducing intracellular calcium.

The second protective mechanism is connected with HSP70 ability to stabilize lysosomal membranes in ischemic cells. During ischemic injury, lysosomal proteases and cathepsins may exit from the lysosome lumen in response to various death signals, such as pro-inflammatory cytokines or oxidative stress. Being released into the cytosol, cathepsins and proteases cause cell autolysis and death. Therefore, by stabilizing lysosomal membranes, HSP70 may limit the exit of dangerous enzymes and prevent cell death induced by cytokines and oxidative stress, the factors that actively contribute to ischemia/reperfusion myocardial injury.

The third defense mechanism is owing to the fact that, in the case of cellular damage, HSP70 translocates to the nucleus and interacts with an ATP-dependent enzyme, poly (ADP-ribose) polymerase. This sensor enzyme detects DNA single-strand breaks and participates in their repair. However, as often happens in biological systems, the same enzyme plays an active role in injury. Whenever the level of cellular oxygen recovers after a period of ischemia, the extent of poly (ADP-ribose) polymerase activation may far exceed the supply levels of available cellular energy needed for such hyperactivation. Hyperactivation of the enzyme in a reperfused heart leads to large amounts of ATP expenditure. As a result, an energy deficit develops, which can damage the cell even more than the oxygen shortage the cells have just experienced. Consequently, it was clearly shown that the interaction of HSP70 with the enzyme limits the excessive activation of poly (ADP-ribose) polymerase and thus preserves ATP stocks in a damaged cell.

The fourth defense mechanism is associated with the ability of HSP70 to inhibit signaling pathways of apoptosis. I have already examined this mechanism in some detail in previous chapters. It is important to add here that, first of all, it was shown that ischemia/reperfusion leads to a significant activation of JNK in cardiomyocytes (Bogoyevitch et al. 1996; Laderoute and Webster 1997), and that activiation level is sufficient to trigger apoptosis in these cells (Wang et al. 1998). Second, experiments on cardiomyocytes showed that HSP70 blocks JNK activation induced by ischemia and reperfusion. This effect of HSP70 was shown to correlate clearly with the increased survival of cells (Gabai et al.1995). Consequently, the protective anti-ischemic effects of HSP70 may be associated with the inhibition of JNK-dependent apoptosis. Thus, experimental and clinical data suggest that intracellular HSP70 protects the myocardium from ischemia/reperfusion injuries during heart surgery (Williams and Benjamin 2000; Lepore et al. 2001; Wang et al. 2003).

8.2.3 Systemic Inflammatory Response Syndrome in Heart Surgery or"Give a Man Enough Rope and He'll Hang Himself"

Apart from ischemia and reperfusion, open heart surgery is always accompanied by development of SIRS. In most cases, it manifests as a temporary tachycardia, arterial hypo- or hypertension, fever without any signs of concomitant infection and slight changes in blood chemistry.

Why does SIRS develop? During heart surgery using the cardiopulmonary bypass, the patient's blood is run by the pump of the heart–lung machine through the tubes of the extracorporeal circulation. These tubes have no inner endothelial layer, as in real blood vessels. Therefore, particular blood cells in contact with the artificial surface are activated. Consequently, after returning to the natural vascular bed of the patient the activated leukocytes release inflammatory cytokines and thus trigger activation of the vascular endothelium. As a result, endothelial cells change their phenotype from the anti-adhesive and anti-inflammatory to the adhesive and proinflammatory type (Kalawski et al. 1998).

The release of various inflammatory mediators by blood cells, activated endothelium and reperfused myocardium into systemic circulation marks the onset

SIRS. Subclinical forms of this syndrome develop in virtually all patients after surgery, even with relatively small lesions. They usually do not pose any problems and are not life-threatening. Moreover, a mild form of systemic inflammation should be regarded as a rather biologically sensible phenomenon. Any open postsurgery wound is an open gate for infection. Therefore, systemic inflammation can be regarded as a form of anticipatory activation of the immune response to potential penetration of pathogenic germs. The systemic inflammatory response syndrome was probably developed and maintained during evolution, so that the body could survive any violation of the body surface integrity, including wounds and burns. Therefore, during surgery the body also uses this ancient form of protection to ensure survival.

However, to prevent excessive inflammation, the body must reliably control every step of the inflammatory process, both locally and at the systemic level. So, in fact, the systemic inflammatory response actually represents a dynamic balance in which the initial hyper-inflammatory phase and the predominance of pro-inflammatory cytokines is followed by a hypo-inflammatory phase with the predominance of anti-inflammatory cytokines (Hiesmayr et al. 1999; Nathan et al. 2000). If the control of the inflammatory processes is disturbed, the syndrome begins to manifest as excessive inflammatory reactions. Among the clinical consequences of an excessive systemic inflammation the most dangerous outcome is the multiple organ failure syndrome (Sablotzki et al. 2002; Rothenburger et al. 2003), which is usually fatal. In some cases, secondary infection contributes to the adverse outcome. However, in many other serious cases, the clinical manifestation of SIRS, although very similar to symptoms of the septic shock syndrome, develops without any symptoms of infection. The SIRS form developing in the absence of infection is called "sepsis-like syndrome". The degree of excessive pro-inflammatory activation of SIRS depends on the reliability of the mechanisms for intracellular, local and systemic regulation of the proinflammatory signaling pathways in immune cells following surgery. A significant component in such a regulation is based on the functions of intracellular and extracellular HSP70.

8.2.4 The Role of HSP70 in the Development and Monitoring of Systemic Inflammatory Response During Heart Surgery: Modus Operandi as a Principle of Biological Regulation

After heart surgery, the extracellular HSP70 level almost always increases. The systemic release of HSP70 has been demonstrated in humans after myocardial infarction (Dybdahl et al. 2005; Satoh et al. 2006) and coronary artery bypass grafting (Dybdahl et al. 2002). The appearance of HSP70 in the systemic circulation results from disturbed cell integrity and the release of intracellular HSP70 from damaged cells into the extracellular space (Saito et al. 2005; Basu et al. 2000). In this case, damaged cells substantially increase HSP70 synthesis to increase HSP70 release at the time of cell necrotic death (Saito et al. 2005). Therefore, the greater the surgical injury, the more HSP70 will be released from cells, and the higher the plasma level of HSP70 (Dybdahl et al. 2004; Cavaillon et al. 2005; Szerafin et al. 2008).

In all cases, the increase in extracellular HSP70 is transient, followed by a more prolonged elevation of circulating proinflammatory cytokines. This transient change in HSP70 and the subsequent increase in cytokine production indicate that extracellular HSP70 may perform signaling functions in triggering the pro-inflammatory component of the systemic inflammatory response. Studies have confirmed that HSP70 does trigger this pro-inflammatory component (Fig. 8.10). In contrast to other intracellular proteins that enter the circulation, extracellular HSP70 interacts with CD14, TLR-2 and/or TLR-4 membrane receptors (Asea et al. 2002; Vabulas et al. 2002; Gross et al. 2003) on the surface of innate immunity cells, such as monocytes, macrophages, natural killer cells and dendritic cells. Intracellular signaling pathways of different immune cell receptors converge on NF-kB. NF-kB activation leads to its translocation from the cytosol to the nucleus, activation of proinflammatory cytokine genes and the development of the innate inflammatory component of a systemic inflammation in a sterile environment.

In this process, innate immunity cells accumulate in areas of inflammation (Tomic et al. 2005; Arispe et al. 2004; Theriault et al. 2005; Vega et al. 2008) in various organs because of the interaction with adhesion molecules on the surface of endothelial cells (Campisi et al. 2003; Chase et al. 2007). In the absence of intracellular damage, the level of intracellular HSP70 in circulating inflammatory cells is low enough. Therefore, HSP70 only slightly inhibits the activity of NF-κB, and so the overall inflammatory balance is pro-inflammatory.

In addition to this, extracellular HSP70 can interact with another group of receptors (see Chap. 5), such as LOX-1, CD91, CD94 and CD40 on the surface of adaptive response cells, such as macrophages, dendritic cells and T cells (Pockley et al. 2008). These ligand-receptor interactions lead to the activation of the adaptive, antigen-nonspecific immune response (Millar et al. 2003; Wang et al. 2006).

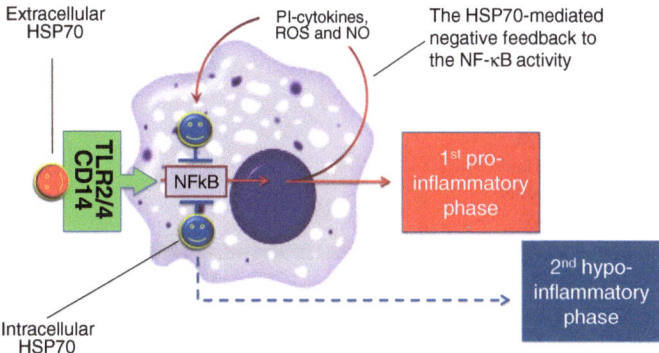

Fig. 8.10 Extracellular HSP70s trigger the pro-inflammatory phase, followed by the delay antiinflammatory phase of the systemic inflammatory response

Thus, extracellular HSP70 may represent a molecular link between myocardial injury and the activation of post-operative, innate, pro-inflammatory and adaptive non-antigen-specific responses of SIRS (Fig. 8.11).

As such, the logic of developing systemic inflammation is as follows: the immune cells activated by extracellular HSP70 start producing large amounts of proinflammatory cytokines, ROS and NO, which, in turn, trigger the synthesis of intracellular HSP70 in immune cells through NFkB (Hamilton et al. 2004; Oehler et al. 2001; Schroder et al. 2003; Temple et al. 2004). This is very important for understanding the switch of systemic inflammatory response syndrome from the pro-inflammatory to the anti-inflammatory direction.

In fact, intracellular HSP70, as distinct from its extracellular partners, inhibits rather than activates NF-κB (Feinstein et al. 1996; Lau et al. 2000), thus blocking the activation of pro-inflammatory cytokine genes. The HSP70-mediated negative feedback to the NF-κB activity is an important immuno-regulatory pathway (Ammirante et al. 2008). The intracellular anti-inflammatory effect of HSP70 is associated with the inhibition of IκB kinase (IKK) activity, which activates NF-kB (Ran et al. 2004). In addition, HSP70 can prevent degradation of the NF-kB inhibitor, I-κBα (Yoo et al. 2000a, b; Weiss et al. 2007) or inhibit translocation of NF-κB from the cytosol to the nucleus (Tang et al. 2007) by physical occlusion of a nuclear pore.

These effects of intracellular HSP70 counteract receptor-mediated proinflammatory signaling, thus preventing the widespread and uncontrolled development of an immune proinflammatory response and ensure multiple organ failure during systemic inflammation.

Therefore, despite the fact that extracellular HSP70 triggers inflammatory responses, which under excessive activation can lead to considerable damage, the inhibition of HSP70 synthesis does not reduce these risks, but leads instead to multiple organ failure and increased mortality (Xiao et al. 1999; Van Molle et al. 2002; Singleton and Wischmeyer 2006). Now it is clear that the inhibition of HSP70 synthesis decreases intracellular levels of HSP70, which depresses the pro-inflammatory signaling cascade.

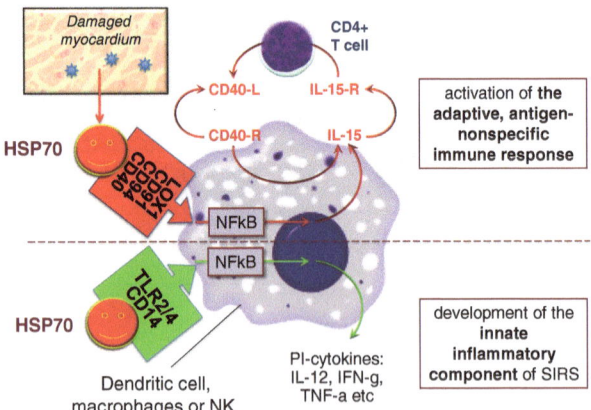

Fig. 8.11 Extracellular HSP70 may represent a molecular link between myocardial injury and the activation of post-operative, innate, pro-inflammatory and adaptive non-antigen-specific responses of the systemic inflammatory response

The effects of intracellular HSP70 have immediately attracted the attention of those pharmacologists and physicians who were concerned about negative and even fatal outcomes in heart patients, who showed excessive activation of the pro-inflammatory component of systemic inflammation after surgery. Moreover, it was already known that HSP70 synthesis can be induced without any cell damage, for example, by the administration of the amino acid glutamine. In addition, it has already been shown that glutamine can increase HSP70 synthesis and reduce damage to lungs during sepsis (Singleton et al. 2005). The pre-operative administration of glutamine in experimental models was shown to increase HSP70 synthesis in different cells and tissues, decrease systemic levels of the proinflammatory cytokines IL-6, IL-8 and NO (Hayashi et al. 2002), and improve post-surgery recovery of the myocardium.

In general, summarizing what you have learned about the role of HSP70 in the regulation of systemic inflammation in heart surgery, we can make the following conclusions.

During systemic inflammation, extracellular HSP70 initiates inflammatory effects through the stimulation of immune cell receptors, whereas intracellular HSP70, in contrast, exerts anti-inflammatory effects on the inflammatory balance of systemic inflammation by inhibiting proinflammatory signaling in immune cells.

Such a situation, when the effects of the same moiety alternatively depend on the site of action, is denoted by the Latin term *modus operandi*. As I have discussed, HSP70 actively uses this principle for the regulation of the systemic inflammatory response syndrome.

8.3 Conclusions

This was the last chapter of the book. I have presented a number of questions about the role of HSP70 in biology and medicine. However, to cover all the important issues is not sufficient in providing a clear answer to all of them. Moreover, I would be dishonest if I did not admit that there remain unresolved features of heat shock proteins function. This is why these proteins remain of interest for basic research and are not yet widely used in the clinical setting.

However, what is definitely known, beyond any doubt, is that the discovery of the molecular basis for regulation and functioning of the entire chaperone system will represent a major breakthrough in discovering the mystery of life. This discovery, in turn, will be essential not only for medicine but for all aspects of biology.

References

Ammirante M, Rosati A, Gentilella A et al (2008) The activity of hsp90 alpha promoter is regulated by NF-kappa B transcription factors. Oncogene 27(8):1175–1178

Ansar S, Burlison JA, Hadden MK et al (2007) A non-toxic Hsp90 inhibitor protects neurons from Abeta-induced toxicity. Bioorg Med Chem Lett 17(7):1984–1990

Archbold RA, Curzen NP (2003) Off-pump coronary artery bypass graft surgery: the incidence of postoperative atrial fibrillation. Heart 89(10):1134–1137

Arispe N, Doh M, Simakova O, Kurganov B, De Maio A (2004) Hsc70 and Hsp70 interact with phosphatidylserine on the surface of PC12 cells resulting in a decrease of viability. FASEB J 18(14):1636–1645

Asea A, Rehli M, Kabingu E et al (2002) Novel signal transduction pathway utilized by extracellular HSP70: role of toll-like receptor (TLR) 2 and TLR4. J Biol Chem 277(17):15028–15034

Basu S, Binder RJ, Suto R, Anderson KM, Srivastava PK (2000) Necrotic but not apoptotic cell death releases heat shock proteins, which deliver a partial maturation signal to dendritic cells and activate the NF-kappa B pathway. Int Immunol 12(11):1539–1546

Bogoyevitch MA, Gillespie-Brown J, Ketterman AJ et al (1996) Stimulation of the stress-activated mitogen-activated protein kinase subfamilies in perfused heart p38/RK mitogen-activated protein kinases and c-Jun N-terminal kinases are activated by ischemia/reperfusion. Circ Res 79(2):162–173

Bonini NM (2002) Chaperoning brain degeneration. Proc Natl Acad Sci U S A 99(Suppl 4):16407–16411

Borchelt DR, Ratovitski T, van Lare J et al (1997) Accelerated amyloid deposition in the brains of transgenic mice coexpressing mutant presenilin 1 and amyloid precursor proteins. Neuron 19:939–945

Brown IR, Gozes I (1998) Stress genes in the nervous system during development and aging diseases. Ann N Y Acad Sci 851:123–128

Calabrese V, Copani A, Testa D et al (2000) Nitric oxide synthase induction in astroglial cell cultures: effect on heat shock protein 70 synthesis and oxidant/antioxidant balance. J Neurosci Res 60(5):613–622

Campion D, Flaman JM, Brice A et al (1995) Mutations of the presenilin I gene in families with early-onset Alzheimer's disease. Hum Mol Genet 4:2373–2377

Campisi J, Leem TH, Fleshner M (2003) Stress-induced extracellular Hsp72 is a functionally significant danger signal to the immune system. Cell Stress Chaperones 8(3):272–286

Caughey B, Lansbury PT (2003) Protofibrils, pores, fibrils, and neurodegeneration: separating the responsible protein aggregates from the innocent bystanders. Annu Rev Neurosci 26:267–298

Cavaillon JM, Adrie C, Fitting C, Adib-Conquy M (2005) Reprogramming of circulatory cells in sepsis and SIRS. J Endotoxin Res 11(5):311–320

Chabrier PE, Demerlé-Pallardy C, Auguet M (1999) Nitric oxide synthases: targets for therapeutic strategies in neurological diseases. Cell Mol Life Sci 55(8–9):1029–1035

Chartier-Harlin MC, Crawford F, Houlden H et al (1991) Early-onset Alzheimer's disease caused by mutations at codon 717 of the beta-amyloid precursor protein gene. Nature 353:844–846

Chase MA, Wheeler DS, Lierl KM, Hughes VS, Wong HR, Page K (2007) Hsp72 induces inflammation and regulates cytokine production in airway epithelium through a TLR4- and NF-kappaB-dependent mechanism. J Immunol 179(9):6318–6324

Chen S, Brown IR (2007) Neuronal expression of constitutive heat shock proteins: implications for neurodegenerative diseases. Cell Stress Chaperones 12:51–58

Christen Y (2000) Oxidative stress and Alzheimer disease. Am J Clin Nutr 71(2):621–629

Cleary JP, Walsh DM, Hofmeister JJ et al (2005) Natural oligomers of the amyloid-beta protein specifically disrupt cognitive function. Nat Neurosci 8:79–84

Clippingdale AB, Wade JD, Barrow CJ (2001) The amyloid-beta peptide and its role in Alzheimer's disease. J Pept Sci 7(5):227–249

Cottrell BA, Galvan V, Banwait S et al (2005) A pilot proteomic study of amyloid precursor interactors in Alzheimer's disease. Ann Neurol 58(2):277–289

Currie RW, Karmazyn M, Kloc M, Mailer K (1998) Heat-shock response is associated with enhanced postischemic ventricular recovery. Circ Res 3:543–549

D'Andrea MR, Nagele RG, Wang HY, Peterson PA, Lee DH (2001) Evidence that neurones accumulating amyloid can undergo lysis to form amyloid plaques in Alzheimer's disease. Histopathology 38(2):120–134

de la Monte SM (2009) Insulin resistance and Alzheimer's disease. BMB Rep 42(8):475–481

Demidov ON, Tyrenko VV, Svistov AS et al (1999) Heat shock proteins in cardiosurgery patients. Eur J Cardiothorac Surg 16(4):444–449

Dewji NN (2005) The structure and functions of the presenilins. Cell Mol Life Sci 62:1109–1119

Dewji NN (2006) Presenilin structure in mechanisms leading to Alzheimer's disease. J Alzheimers Dis 10:277–290

Dewji NN, Do C, Bayney RM (1995) Transcriptional activation of Alzheimer's beta-amyloid precursor protein gene by stress. Brain Res Mol Brain 33:245–253

Dickey CA, Dunmore J, Lu B et al (2006) HSP induction mediates selective clearance of tau phosphorylated at proline-directed Ser/Thr sites but not KXGS (MARK) sites. FASEB J 20(6):753–755

Dickey CA, Koren J, Zhang YJ et al (2008) (2008) Akt and CHIP coregulate tau degradation through coordinated interactions. Proc Natl Acad Sci U S A. 105(9):3622–3627

Dou F, Netzer WJ, Tanemura K et al (2003) Chaperones increase association of tau protein with microtubules. Proc Natl Acad Sci U S A 100(2):721–726

Duff K, Eckman C, Zehr C et al (1996) Increased amyloid-beta42(43) in brains of mice expressing mutant presenilin 1. Nature 383:710–713

Dybdahl B, Slørdahl SA, Waage A, Kierulf P, Espevik T, Sundan A (2005) Myocardial ischaemia and the inflammatory response: release of heat shock protein 70 after myocardial infarction. Heart 91(3):299–304

Dybdahl B, Wahba A, Haaverstad R et al (2004) On-pump versus off-pump coronary artery bypass grafting: more heat-shock protein 70 is released after on-pump surgery. Eur J Cardiothorac Surg 25(6):985–992

Dybdahl B, Wahba A, Lien E et al (2002) Inflammatory response after open heart surgery: release of heat-shock protein 70 and signaling through toll-like receptor-4. Circulation 105(6):685–690

Elliott E, Tsvetkov P, Ginzburg I (2007) BAG-1 associates with Hsc70.Tau complex and regulates the proteasomal degradation of Tau protein. J Biol Chem 51:37276–37284

Estus S, Golde TE, Kunishita T et al (1992) Potentially amyloidogenic, carboxyl-terminal derivatives of the amyloid protein precursor. Science 255:726–728

Evans CG, Wisén S, Gestwicki JE (2006) Heat shock proteins 70 and 90 inhibit early stages of amyloid beta-(1–42) aggregation in vitro. J Biol Chem 281(44):33182–33191

Feinstein DL, Galea E, Aquino DA, Li GC, Xu H, Reis DJ (1996) Heat shock protein 70 suppresses astroglial-inducible nitric-oxide synthase expression by decreasing NFkappaB activation. J Biol Chem 271(30):17724–17732

Fonte J, Bates KA, Robertson TA, Martins RN, Harvey AR (2002) Chronic gliosis triggers Alzheimer's disease-like processing of amyloid precursor protein. Neuroscience 113(4):785–796

Gabai VL, Kabakov AE (1993) Rise in heat-shock protein level confers tolerance to energy deprivation. FEBS Lett 327(3):247–250

Gabai VL, Mosina VA, Budagova KR, Kabakov AE (1995) Spontaneous overexpression of heat-shock proteins in Ehrlich ascites carcinoma cells during in vivo growth. Biochem Mol Biol Int 35(1):95–102

Getchell TV, Kulkarni-Narla A, Schmitt FA, Getchell ML (1996) Manganese and copper-zinc superoxide dismutases in the human olfactory mucosa: increased immunoreactivity in Alzheimer's disease. Exp Neurol 140(2):115–125

Giannessi D, Caselli C, Vitale RL et al (2003) Possible cardioprotective effect of heat shock proteins during cardiac surgery in pediatric patients. Pharmacol Res 48(5):519–529

Goate A, Chartier-Harlin MC, Mullan M et al (1991) Segregation of a missense mutation in the amyloid precursor protein gene with familial Alzheimer's disease. Nature 349:704–706

Golde TE, Estus S, Younkin LH, Selkoe DJ, Younkin SG (1992) Processing of the amyloid protein precursor to potentially amyloidogenic derivatives. Science 255:728–730

Gong Y, Chang L, Viola KL et al (2003) Alzheimer's disease-affected brain: presence of oligomeric A beta ligands (ADDLs) suggests a molecular basis for reversible memory loss. Proc Natl Acad Sci USA 100:10417–10422

Gouras GK, Tsai J, Naslund J et al (2000) Intraneuronal Abeta42 accumulation in human brain. Am J Pathol 156(1):15–20

Gross C, Hansch D, Gastpar R, Multhoff G (2003) Interaction of heat shock protein 70 peptide with NK cells involves the NK receptor CD94. Biol Chem 384(2):267–279

Guillozet-Bongaarts AL, Garcia-Sierra F, Reynolds MR et al (2005) Tau truncation during neurofibrillary tangle evolution in Alzheimer's disease. Neurobiol Aging 26(7):1015–1022

Halverson K, Fraser PE, Kirschner DA, Lansbury PT Jr (1990) Molecular determinants of amyloid deposition in Alzheimer's disease: conformational studies of synthetic beta-protein fragments. Biochem 29:2639–2644

Hamilton KL, Gupta S, Knowlton AA (2004) Estrogen and regulation of heat shock protein expression in female cardiomyocytes: cross-talk with NF kappa B signaling. J Mol Cell Cardiol 36(4):577–584

Hayashi Y, Sawa Y, Fukuyama N, Nakazawa H, Matsuda H (2002) Preoperative glutamine administration induces heat-shock protein 70 expression and attenuates cardiopulmonary bypass-induced inflammatory response by regulating nitric oxide synthase activity. Circulation 106(20):2601–2607

Hiesmayr MJ, Spittler A, Lassnigg A et al (1999) Alterations in the number of circulating leucocytes, phenotype of monocyte and cytokine production in patients undergoing cardiothoracic surgery. Clin Exp Immunol 2:315–323

Hock C, Konietzko U, Streffer JR et al (2003) Papassotiropoulos A, Nitsch RM. Antibodies against beta-amyloid slow cognitive decline in Alzheimer's disease. Neuron 38:547–554

Holcomb L, Gordon MN, McGowan E et al (1998) Accelerated Alzheimer-type phenotype in transgenic mice carrying both mutant amyloid precursor protein and presenilin 1 transgenes. Nat Med 4:97–100

Holcomb LA, Gordon MN, Jantzen P, Hsiao K, Duff K, Morgan D (1999) Behavioral changes in transgenic mice expressing both amyloid precursor protein and presenilin-1 mutations: lack of association with amyloid deposits. Behav Genet 29:177–185

Hsiao K, Chapman P, Nilsen S et al (1996) Correlative memory deficits, Abeta elevation, and amyloid plaques in transgenic mice. Science 274:99–102

Hu J, LaDu MJ, Van Eldik LJ (1998) Apolipoprotein E attenuates beta-amyloid-induced astrocyte activation. J Neurochem 71(4):1626–1634

Hutton M, Lendon CL, Rizzu P et al (1998) Association of missense and 5′-splice-site mutations in tau with the inherited dementia FTDP-17. Nature 393:702–705

Jayakumar J, Suzuki K, Khan M et al (2000) Gene therapy for myocardial protection: transfection of donor hearts with heat shock protein 70 gene protects cardiac function against ischemia-reperfusion injury. Circulation 102(19 Suppl 3):III302-6

Jayakumar J, Suzuki K, Sammut IA et al (2001) Heat shock protein 70 gene transfection protects mitochondrial and ventricular function against ischemia-reperfusion injury. Circulation 104(12 Suppl 1):303–307

Johnson G, Refolo LM, Wallace W (1993) Heat-shocked neuronal PC12 cells reveal Alzheimer's disease–associated alterations in amyloid precursor protein and tau. Ann N Y Acad Sci 695:194–197

Kabakov AE, Gabai VL (1995) Heat shock-induced accumulation of 70-kDa stress protein (HSP70) can protect ATP-depleted tumor cells from necrosis. Exp Cell Res 217(1):15–21

Kakimura J, Kitamura Y, Takata K et al (2002) Possible involvement of ER chaperone Grp78 on reduced formation of amyloid-beta deposits. Ann N Y Acad Sci 977:327–332

Kalawski R, Bugajski P, Smielecki J et al (1998) Soluble adhesion molecules in reperfusion during coronary bypass grafting. Eur J Cardiothorac Surg 14(3):290–295

Kirby BA, Merril CR, Ghanbari H, Wallace WC (1994) Heat shock proteins protect against stress-related phosphorylation of tau in neuronal PC12 cells that have acquired thermotolerance. J Neurosci 14(9):5687–5693

Klucken J, Shin Y, Masliah E, Hyman BT, McLean PJ (2004) Hsp70 Reduces alpha-Synuclein Aggregation and Toxicity. J Biol Chem 279(24):25497–25502

Koren J 3rd, Jinwal UK, Lee DC et al (2009) Chaperone signalling complexes in Alzheimer's disease. J Cell Mol Med 13(4):619–630

Kosik KS, Ahn J, Stein R, Yeh LA (2002) Discovery of compounds that will prevent tau pathology. J Mol Neurosci 19(3):261–266

Kumar-Singh S, De Jonghe C, Cruts M et al (2000) Nonfibrillar diffuse amyloid deposition due to a gamma (42)-secretase site mutation points to an essential role for N-truncated A beta(42) in Alzheimer's disease. Hum Mol Genet 18:2589–2598

Laderoute KR, Webster KA (1997) Hypoxia/reoxygenation stimulates Jun kinase activity through redox signaling in cardiac myocytes. Circ Res 80(3):336–344

Lau SS, Griffin TM, Mestril R (2000) Protection against endotoxemia by HSP70 in rodent cardiomyocytes. Am J Physiol Heart Circ Physiol 278(5):1439–1445

Lepore DA, Knight KR, Anderson RL, Morrison WA (2001) Role of priming stresses and Hsp70 in protection from ischemia-reperfusion injury in cardiac and skeletal muscle. Cell Stress Chaperones 6(2):93–96

Lester-Coll N, Rivera EJ, Soscia SJ, Doiron K, Wands JR, de la Monte SM (2006) Intracerebral streptozotocin model of type 3 diabetes: relevance to sporadic Alzheimer's disease. J Alzheimers Dis 9(1):13–33

Magrané J, Smith RC, Walsh K, Querfurth HW (2004) Heat shock protein 70 partici- pates in the neuroprotective response to intracellularly expressed beta-amyloid in neurons. J Neurosci 24:1700–1706

Marucci A, Miscio G, Padovano L et al (2009) The role of HSP70 on ENPP1 expression and insulin-receptor activation. J Mol Med (Berl) 87(2):139–144

McCarty MF (2005) Induction of heat shock proteins may combat insulin resistance. Med Hypotheses 66(3):527–534

Meyer-Luehmann M, Spires-Jones TL, Prada C et al (2008) Rapid appearance and local toxicity of amyloid-beta plaques in a mouse model of Alzheimer's disease. Nature 451:720–724

Millar DG, Garza KM, Odermatt B, Elford AR, Ono N, Li Z, Ohashi PS (2003) Hsp70 promotes antigen-presenting cell function and converts T-cell tolerance to autoimmunity in vivo. Nat Med 9(12):1469–1476

Morino S, Kondo T, Sasaki K et al (2008) Mild electrical stimulation with heat shock ameliorates insulin resistance via enhanced insulin signaling. PLoS ONE 3(12):4068

Mullan M, Houlden H, Windelspecht M et al (1992) A locus for familial early-onset Alzheimer's disease on the long arm of chromosome 14, proximal to the alpha 1-antichymotrypsin gene. Nat Genet 2:340–342

Nathan N, Preux PM, Feiss P, Denizot Y (2000) Plasma interleukin-4, interleukin-10, and interleukin-13 concentrations and complications after coronary artery bypass graft surgery. J Cardiothorac Vasc Anesth 14(2):156–160

Nitsch RM, Hock C (2008) Targeting beta-amyloid pathology in Alzheimer's Disease with Abeta immunotherapy. Neurotherapeutics 5:415–420

Noble W, Planel E, Zehr C et al (2005) Inhibition of glycogen synthase kinase-3 by lithium correlates with reduced tauopathy and degeneration in vivo. Proc Natl Acad Sci U S A 102(19):6990–6995

Oehler R, Pusch E, Zellner M et al (2001) Cell type-specific variations in the induction of hsp70 in human leukocytes by fever like whole body hyperthermia. Cell Stress Chaperones 6(4):306–315

Ohtsuka K, Suzuki T (2000) Roles of molecular chaperones in the nervous system. Brain Res Bull 53(2):141–146

Okubo S, Wildner O, Shah MR, Chelliah JC, Hess ML, Kukreja RC (2001) Gene transfer of heat-shock protein 70 reduces infarct size in vivo after ischemia/reperfusion in the rabbit heart. Circulation 103(6):877–881

Opar A (2008) Mixed results for disease-modification strategies for Alzheimer's disease. Nat Rev Drug Discov 7(9):717–718

Osborne R (2008) Myriad stumbles, Wyeth closes on Alzheimer's. Nat Biotechnol 26(8):841–843

Pappolla MA, Sos M, Omar RA, Sambamurti K (1996) The heat shock/oxidative stress connection. Relevance to Alzheimer disease. Mol Chem Neuropathol 28(1–3):21–34

Pei JJ, Khatoon S, An WL, Nordlinder M et al (2003) Role of protein kinase B in Alzheimer's neurofibrillary pathology. Acta Neuropathol 105(4):381–392

Perez-Tur J, Croxton R, Wright K et al (1996) A further presenilin 1 mutation in the exon 8 cluster in familial Alzheimer's disease. Neurodegeneration 5:207–212

Petrucelli L, Dickson D, Kehoe K et al (2004) CHIP and Hsp70 regulate tau ubiquitination, degradation and aggregation. Hum Mol Genet 13(7):703–714

Piccoletti R, Schiaffonati L, Maroni P, Bendinelli P, Tiberio L (2001) Hyperthermia induces gene expression of heat shock protein 70 and phosphorylation of mitogen activated protein kinases in the rat cerebellum. Neurosci Lett 312(2):75–78

Pickering-Brown S, Baker M, Yen SH et al (2000) Pick's disease is associated with mutations in the tau gene. Ann Neurol 48(6):859–867

Pockley AG, Muthana M, Calderwood SK (2008) The dual immunoregulatory roles of stress proteins. Trends Biochem Sci 33(2):71–79

Ramsden M, Kotilinek L, Forster C et al (2005) Age-dependent neurofibrillary tangle formation, neuron loss, and memory impairment in a mouse model of human tauopathy (P301L). J Neurosci 25(46):10637–10647

Ran R, Lu A, Zhang L et al (2004) Hsp70 promotes TNF-mediated apoptosis by binding IKK gamma and impairing NF-kappa B survival signaling. Genes Dev 18(12):1466–1481

Reger MA, Watson GS, Green PS et al (2008) Intranasal insulin improves cognition and modulates β-amyloid in early AD. Neurology 70(6):440–448

Renkawek K, Bosman GJ, de Jong WW (1994) Expression of small heat-shock protein hsp 27 in reactive gliosis in Alzheimer disease and other types of dementia. Acta Neuropathol 87(5):511–519

Renkawek K, Bosman GJ, Gaestel M (1993) Increased expression of heat-shock protein 27 kDa in Alzheimer disease: a preliminary study. Neuroreport 5(1):14–16

Rizzini C, Goedert M, Hodges JR et al (2000) Tau gene mutation K257T causes a tauopathy similar to Pick's disease. J Neuropathol Exp Neurol 59:990–1001

Rothenburger M, Tjan TD, Schneider M et al (2003) The impact of the pro- and anti-inflammatory immune response on ventilation time after cardiac surgery. Cytometry B Clin Cytom 53(1):70–74

Sablotzki A, Friedrich I, Mühling J et al (2002) The systemic inflammatory response syndrome following cardiac surgery: different expression of proinflammatory cytokines and procalcitonin in patients with and without multiorgan dysfunctions. Perfusion 17(2):103–109

Sadik K, Wilcock G (2003) The increasing burden of Alzheimer disease. Alzheimer Dis Assoc Disord 17(Suppl 3):75–79

Saito K, Dai Y, Ohtsuka K (2005) Enhanced expression of heat shock proteins in gradually dying cells and their release from necrotically dead cells. Exp Cell Res 310(1):229–236

Salbaum JM, Weidemann A, Lemaire HG, Masters CL, Beyreuther K (1998) The promoter of Alzheimer's disease amyloid A4 precursor gene. EMBO J 7(9):2807–2813

Santacruz K, Lewis J, Spires T et al (2005) Tau suppression in a neurodegenerative mouse model improves memory function. Science 309(5733):476–481

Sarkar M, Kuret J, Lee G (2008) Two motifs within the tau microtubule-binding domain mediate its association with the hsc70 molecular chaperone. J Neurosci Res 86(12):2763–2773

Sasaki A, Yamaguchi H, Ogawa A, Sugihara S, Nakazato Y (1997) Microglial activation in early stages of amyloid beta protein deposition. Acta Neuropathol 94(4):316–322

Satoh M, Shimoda Y, Akatsu T, Ishikawa Y, Minami Y, Nakamura M (2006) Elevated circulating levels of heat shock protein 70 are related to systemic inflammatory reaction through monocyte Toll signal in patients with heart failure after acute myocardial infarction. Eur J Heart Fail 8(8):810–815

Schenk DB, Seubert P, Grundman M, Black R (2005) A beta immunotherapy: lessons learned for potential treatment of Alzheimer's disease. Neurodegener Dis 2:255–260

Schröder O, Schulte KM, Ostermann P et al (2003) Heat shock protein 70 genotypes HSPA1B and HSPA1L influence cytokine concentrations and interfere with outcome after major injury. Crit Care Med 31(1):73–79

Shankar GM, Bloodgood BL, Townsend M et al (2007) Natural oligomers of the Alzheimer amyloid-beta protein induce reversible synapse loss by modulating an NMDA-type glutamate receptor-dependent signaling pathway. J Neurosci 27:2866–2875

Shankar GM, Li S, Mehta TH et al (2008) Amyloid-beta protein dimers isolated directly from Alzheimer's brains impair synaptic plasticity and memory. Nat Med 14(8):837–842

Sherrington R, Rogaev EI, Liang Y et al (1995) Cloning of a gene bearing missense mutations in early-onset familial Alzheimer's disease. Nature 375:754–760

Shimura H, Miura-Shimura Y, Kosik KS (2004) Binding of tau to heat shock protein 27 leads to decreased concentration of hyperphosphorylated tau and enhanced cell survival. J Biol Chem 279(17):17957–17962

Shoji M, Golde TE, Ghiso J et al (1992) Production of the Alzheimer amyloid beta protein by normal proteolytic processing. Science 258:126–129

Simard AR, Soulet D, Gowing G, Julien JP, Rivest S (2006) Bone marrow-derived microglia play a critical role in restricting senile plaque formation in Alzheimer's disease. Neuron 49(4):489–502

Singleton KD, Serkova N, Beckey VE, Wischmeyer PE (2005) Glutamine attenuates lung injury and improves survival after sepsis: role of enhanced heat shock protein expression. Crit Care Med 33(6):1206–1213

Singleton KD, Wischmeyer PE (2006) Effects of HSP70.1/3 gene knockout on acute respiratory distress syndrome and the inflammatory response following sepsis. Am J Physiol Lung Cell Mol Physiol 290(5):956–961

Smith RC, Rosen KM, Pola R, Magrané J (2005) Stress proteins in Alzheimer's disease. Int J Hyperthermia 21(5):421–431

Spillantini MG, Murrell JR, Goedert M, Farlow MR, Klug A, Ghetti B (1998) Mutation in the tau gene in familial multiple system tauopathy with presenile dementia. Proc Natl Acad Sci USA 95:7737–7741

Spires-Jones TL, de Calignon A, Matsui T et al (2008) In vivo imaging reveals dissociation between caspase activation and acute neuronal death in tangle-bearing neurons. J Neurosci 28:862–867

St Rammos K, Koullias GJ, Hassan MO et al (2002) Low preoperative HSP70 atrial myocardial levels correlate significantly with high incidence of postoperative atrial fibrillation after cardiac surgery. Cardiovasc Surg 10(3):228–232

Stalder AK, Ermini F, Bondolfi L et al (2005) Invasion of hematopoietic cells into the brain of amyloid precursor protein transgenic mice. J Neurosci 25(48):11125–11132

Sun YX, Wright HT, Janciauskiene S (2002) Glioma cell activation by Alzheimer's peptide Abeta1-42, alpha1-antichymotrypsin, and their mixture. Cell Mol Life Sci 59(10):1734–1743

Szerafin T, Hoetzenecker K, Hacker S et al (2008) Heat shock proteins 27, 60, 70, 90alpha, and 20S proteasome in on-pump versus off-pump coronary artery bypass graft patients. Ann Thorac Surg 85(1):80–87

Taggart DP, Bakkenist CJ, Biddolph SC, Graham AK, McGee JO (1997) Induction of myocardial heat shock protein 70 during cardiac surgery. J Pathol 182(3):362–366

Takeuchi A, Irizarry MC, Duff K et al (2000) Age-related amyloid beta deposition in transgenic mice overexpressing both Alzheimer mutant presenilin 1 and amyloid beta precursor protein Swedish mutant is not associated with global neuronal loss. Am J Pathol 157:331–339

Tang D, Kang R, Xiao W, Wang H, Calderwood SK, Xiao X (2007) The anti-inflammatory effects of heat shock protein 72 involve inhibition of high-mobility-group box 1 release and proinflammatory function in macrophages. J Immunol 179(2):1236–1244

Temple SE, Cheong KY, Ardlie KG, Sayer D, Waterer GW (2004) The septic shock associated HSPA1B1267 polymorphism influences production of HSPA1A and HSPA1B. Intensive Care Med 30(9):1761–1767

Thériault JR, Mambula SS, Sawamura T, Stevenson MA, Calderwood SK (2005) Extracellular HSP70 binding to surface receptors present on antigen presenting cells and endothelial/epithelial cells. FEBS Lett 579(9):1951–1960

Tomic V, Russwurm S, Möller E et al (2005) Transcriptomic and proteomic patterns of systemic inflammation in on-pump and off-pump coronary artery bypass grafting. Circulation 112(19):2912–2920

Tytell M, Greenberg SG, Lasek RJ (1986) Heat shock-like protein is transferred from glia to axon. Brain Res 363:161–164

Vabulas RM, Ahmad-Nejad P, Ghose S et al (2002) HSP70 as endogenous stimulus of the Toll/interleukin-1 receptor signal pathway. J Biol Chem 277(17):15107–15112

Van Broeckhoven C, Haan J, Bakker E et al (1990) Amyloid beta protein precursor gene and hereditary cerebral hemorrhage with amyloidosis (Dutch). Science 248:1120–1122

Van Molle W, Wielockx B, Mahieu T et al (2002) HSP70 protects against TNF-induced lethal inflammatory shock. Immunity 16(5):685–695

Vega VL, Rodríguez-Silva M, Frey T et al (2008) Hsp70 translocates into the plasma membrane after stress and is released into the extracellular environment in a membrane-associated form that activates macrophages. J Immunol 180(6):4299–4307

Vincent SR, Kimura H (1992) Histochemical mapping of nitric oxide synthase in the rat brain. Neuroscience 46(4):755–784

Wakutani Y, Urakami K, Shimomura T, Takahashi K (1995) Heat shock protein 70 mRNA levels in mononuclear blood cells from patients with dementia of the Alzheimer type. Dementia 6(6):301–305

Walsh DM, Klyubin I, Fadeeva JV et al (2002) Naturally secreted oligomers of amyloid beta protein potently inhibit hippocampal long-term potentiation in vivo. Nature 416:535–539

Wang R, Kovalchin JT, Muhlenkamp P, Chandawarkar RY (2006) Exogenous heat shock protein 70 binds macrophage lipid raft microdomain and stimulates phagocytosis, processing, and MHC-II presentation of antigens. Blood 107(4):1636–1642

Wang S, Diller KR, Aggarwal SJ (2003) Kinetics study of endogenous heat shock protein 70 expression. J Biomech Eng 125(6):794–797

Wang Y, Su B, Sah VP, Brown JH, Han J, Chien KR (1998) Cardiac hypertrophy induced by mitogen-activated protein kinase kinase 7, a specific activator for c-Jun NH2-terminal kinase in ventricular muscle cells. J Biol Chem 273(10):5423–5426

Weaver CL, Espinoza M, Kress Y, Davies P (2000) Conformational change as one of the earliest alterations of tau in Alzheimer's disease. Neurobiol Aging 21(5):719–727

Weiss YG, Bromberg Z, Raj N et al (2007) Enhanced heat shock protein 70 expression alters proteasomal degradation of IkappaB kinase in experimental acute respiratory distress syndrome. Crit Care Med 35(9):2128–2138

Wilcock DM, Alamed J, Gottschall PE et al (2006) Deglycosylated anti-amyloid-beta antibodies eliminate cognitive deficits and reduce parenchymal amyloid with minimal vascular consequences in aged amyloid precursor protein transgenic mice. J Neurosci 26:5340–5346

Wilcock DM, Rojiani A, Rosenthal A et al (2004) Passive immunotherapy against Abeta in aged APP-transgenic mice reverses cognitive deficits and depletes parenchymal amyloid deposits in spite of increased vascular amyloid and microhemorrhage. J Neuroinflammation 1:24

Williams RS, Benjamin IJ (2000) Protective responses in the ischemic myocardium. J Clin Invest 106(7):813–818

Wirths O, Multhaup G, Czech C et al (2001) Reelin in plaques of beta-amyloid precursor protein and presenilin-1 double-transgenic mice. Neurosci Lett 316(3):145–148

Xiao X, Zuo X, Davis AA et al (1999) HSF1 is required for extra-embryonic development, postnatal growth and protection during inflammatory responses in mice. EMBO J 18(21):5943–5952

Yan SD, Fu J, Soto C et al (1997) An intracellular protein that binds amyloid-beta peptide and mediates neurotoxicity in Alzheimer's disease. Nature 389(6652):689–695

Yang SN, Hsieh WY, Liu DD, Tsai LM, Tung CS, Wu JN (1998) The involvement of nitric oxide in synergistic neuronal damage induced by beta-amyloid peptide and glutamate in primary rat cortical neurons. Chin J Physiol 41(3):175–179

Yoo BC, Seidl R, Cairns N, Lubec G (1999) Heat-shock protein 70 levels in brain of patients with Down syndrome and Alzheimer's disease. J Neural Transm Suppl 57:315–322

Yoo CG, Lee S, Lee CT, Kim YW, Han SK, Shim YS (2000a) Anti-inflammatory effect of heat shock protein induction is related to stabilization of I kappa B alpha through preventing I kappa B kinase activation in respiratory epithelial cells. J Immunol 164(10):5416–5423

Yoo JC, Pae HO, Choi BM et al (2000b) Lonizing radiation potentiates the induction of nitric oxide synthase by interferon-gamma (Ifn-gamma) or Ifn-gamma and lipopolysaccharide in bnl cl.2 murine embryonic liver cells: role of hydrogen peroxide. Free Radic Biol Med 28(3):390–396

Zachayus JL, Benatmane S, Plas C (1996) Role of Hsp70 synthesis in the fate of the insulin-receptor complex after heat shock in cultured fetal hepatocytes. J Cell Biochem 61(2):216–229

Index

I. Malyshev, *Immunity, Tumors and Aging: The Role of HSP70*,   141
SpringerBriefs in Biochemistry and Molecular Biology,
DOI: 10.1007/978-94-007-5943-5, © The Author(s) 2013